Online Portfolio Selection

Principles and Algorithms

T0304148

Online Portfolio Selection

Principles and Algorithms

Bin Li and Steven C.H. Hoi

CRC Press
Taylor & Francis Group
Boca Raton London New York

CRC Press is an imprint of the
Taylor & Francis Group, an **informa** business

First published in hardback 2020
First published in paperback 2024

Published 2016 by CRC Press
2385 NW Executive Center Drive, Suite 320, Boca Raton FL 33431

and by CRC Press
4 Park Square, Milton Park, Abingdon, Oxon, OX14 4RN

CRC Press is an imprint of Taylor & Francis Group, LLC

© 2016, 2020, 2024 Taylor & Francis Group, LLC

ISBN: 978-1-4822-4963-7 (hbk)
ISBN: 978-1-138-89410-5 (pbk)
ISBN: 978-1-351-22918-0 (ebk)

DOI: 10.1201/b19011

Contents

List of Figures

List of Tables

List of Notations

α	Decaying factor for MAR
$\bar{x}_t$	Arithmetic mean of t-th (predict or realized) price relatives
$\odot$	Element-wise product of two vectors
Δ_m	Simplex domain
$\ell_\epsilon(\mathbf{b}; \mathbf{x}_t)$	ϵ-insensitive loss function
ϵ	Sensitivity parameter for PAMR
γ	Proportional transaction cost rate
$\int$	Integral
$\mathbb{R}_+^m$	Domain of m-dimensional vectors with positive real elements
$\mathbf{1}$	Vector of all 1
$\mathbf{b} \cdot \mathbf{x}$	Dot product of vector $\mathbf{b}$ and vector $\mathbf{x}$
$\mathbf{b}^\top \mathbf{x}$	Dot product of vector $\mathbf{b}$ and vector $\mathbf{x}$
$\mathbf{b}_1^n$	A portfolio strategy for n periods
$\mathbf{b}_t$	Portfolio vector for t-th period
$\mathbf{I}$	Identity matrix
$\mathbf{p}_t$	Vector of closing prices for t-th period
$\mathbf{x}_1^n$	Whole market windows for n periods
$\mathbf{x}_s^e$	Market windows from period s to e
$\mathbf{x}_t$	Vector of price relatives (simple gross return) for t-th period
$\mathcal{N}(\mu, \Sigma)$	Normal distribution with mean μ and covariance matrix Σ
$\det(\Sigma)$	Determinant of matrix Σ
$\text{EMA}_t(\alpha)$	Exponential moving average
$\text{regret}_n(Alg)$	Regret of Alg strategy with respect to BCRP
$\text{SMA}_t(w)$	Simple moving average
$\mu(\cdot)$	A distribution on the space of valid portfolio
ϕ	Confidence parameter
$\Phi(\cdot)$	Cumulative function of normal distribution
ρ	Correlation coefficient threshold
$\tilde{\mathbf{x}}_{t+1}$	Predicted price relative vector
$C_t(w, \rho)$	Correlation-similar set
i	Index of an asset
m	The number of assets

n	The number of trading periods
P	Maximum correlation coefficient
S_t	Cumulative return till the end of t-th period
s_t	Daily return for t-th period
$S_t(Alg)$	Cumulative return achieved by Alg until the end of t-th period
t	Index of a trading period
W	Maximum window size
w	Window size
$W_t(Alg)$	Exponential growth rate achieved by Alg until the t-th period
C	Aggressive parameter for PAMR
CORN(w, ρ)	Correlation-driven nonparametric expert
CORN-K	CORN algorithm with top-K aggregation
CORN-U	CORN algorithm with uniform aggregation

Preface

Introduction

Computational intelligence techniques, including machine learning and data mining, have significantly reshaped the financial investment community over recent decades. Examples include high-frequency trading and algorithmic trading. This book studies a fundamental problem in computational finance, or online portfolio selection (OLPS), which aims to sequentially determine optimal allocations across a set of assets. This book investigates this problem by conducting a comprehensive survey on existing principles and presenting a family of new strategies using machine-learning techniques. A back-test system using historical data has been developed to evaluate the performance of trading strategies.

Our goal in writing this monograph is to present a self-contained text to a wide range of audiences, including graduate students in finance, computer science, and statistics, as well as researchers and engineers who are interested in computational investment. The readers are encouraged to visit our project website for more updates: http://olps.stevenhoi.org

Organization

Part I introduces the OLPS problem. Chapter 1 introduces the background and summarizes the contributions of this book. Chapter 2 formally formulates OLPS as a sequential decision task.

Part II presents some key principles for this task. Chapter 3 summarizes three benchmarks: the Buy-and-Hold strategy, Best Stock strategy, and Constant Rebalanced Portfolios. Chapter 4 presents the principle of Follow the Winner, which moves weights from winning assets to losing assets. Chapter 5 presents an opposite principle called Follow the Loser, which moves weights from losers to winners. Chapter 6 demonstrates the principle of Pattern Matching, which exploits similar patterns among historical markets. Chapter 7 talks about Meta-Learning, which views the strategies as assets, and thus hyperstrategies.

Part III designs four novel algorithms to solve the OLPS problem. All algorithms apply the state-of-the-art machine-learning techniques to the task. Chapter 8 designs a new strategy named CORrelation-driven Nonparametric (CORN) learning, which overcomes the limitations of existing pattern matching–based strategies using Euclidean distance to measure the similarity between two patterns. Chapter 9 develops Passive–Aggressive Mean Reversion (PAMR), which is based on the first-order passive–aggressive online learning method, and Chapter 10 designs

Confidence-Weighted Mean Reversion (CWMR), which is based on the second-order confidence-weighted online learning method. Chapter 11 assumes multiple-period mean reversion, or so-called Moving Average Reversion (MAR), and presents a new OLPS strategy named Online Moving Average Reversion (OLMAR), which exploits MAR by applying online learning techniques.

Part IV presents empirical studies for benchmarking the performance of the proposed algorithms. Chapter 12 discusses issues related to the implementation of a back-test system, which is widely used in the evaluation of trading strategies. Chapter 13 shows the empirical results on six historical markets. Our empirical results show that (i) the proposed algorithms generally outperform the state of the art in terms of the cumulative return and risk-adjusted return and (ii) the proposed algorithms are highly efficient and scalable for large-scale OLPS in real-world applications. Chapter 14 discusses various assumptions during the study.

Part V concludes the book and presents some potential future directions.

MATLAB® is a registered trademark of The MathWorks, Inc. For product information, please contact:

The MathWorks, Inc.
3 Apple Hill Drive
Natick, MA 01760-2098, USA
Tel: 508-647-7000
Fax: 508-647-7001
E-mail: info@mathworks.com
Web: www.mathworks.com

Acknowledgments

We thank Peilin Zhao, Doyen Sahoo, Vivekanand Gopalkrishnan, and Dingjiang Huang, who had participated in different stages of our online portfolio selection project and made different contributions to some of our related research in this area, which are the core foundations of this book. We gratefully acknowledge research support from both Wuhan University and Singapore Management University. This book was supported in part by the National Natural Science Foundation of China (71401128); the Scientific Research Foundation for the Returned Overseas Chinese Scholars, State Education Ministry; the Fundamental Research Funds for the Central Universities; and the MOE tier-1 research grant from Singapore Management University. Special thanks also goes to Nanyang Technological University, where the authors started the early initiatives of this project and completed this project.

Bin Li
Economics and Management School
Wuhan University, People's Republic of China

Steven C.H. Hoi
School of Information Systems
Singapore Management University, Singapore

Authors

Dr. Bin Li received a bachelor's degree in computer science from Huazhong University of Science and Technology, Wuhan, China, and a bachelor's degree in economics from Wuhan University, Wuhan, China, in 2006. He earned a PhD degree from the School of Computer Engineering of Nanyang Technological University, Singapore, in 2013. He completed the CFA Program in 2013. He is currently an associate professor of finance at the Economics and Management School of Wuhan University. Dr. Li was a postdoctoral research fellow at the Nanyang Business School of Nanyang Technological University. His research interests are computational finance and machine learning. He has published several academic papers in premier conferences and journals.

Dr. Steven C.H. Hoi received his bachelor's degree in computer science from Tsinghua University, Beijing, China, in 2002, and both his master's and PhD degrees in computer science and engineering from The Chinese University of Hong Kong, Hong Kong, China, in 2004 and 2006, respectively. He is currently an associate professor in the School of Information Systems, Singapore Management University, Singapore. Prior to joining SMU, he was a tenured associate professor in the School of Computer Engineering, Nanyang Technological University, Singapore. His research interests are machine learning and data mining and their applications to tackle real-world big data challenges across varied domains, including computational finance, multimedia information retrieval, social media, web search and data mining, computer vision and pattern recognition, and so on. Dr. Hoi has published more than 150 referred articles in premier international journals and conferences. As an active researcher in his research communities, he has served as general co-chair for ACM SIGMM Workshops on Social Media (WSM'09-11), program co-chair for Asian Conference on Machine Learning (ACML'12), editor for *Social Media Modeling and Computing*, guest editor for journals such as *Machine Learning* and *ACM TIST*, associate editor-in-chief of *Neurocomputing*, associate editor for several reputable journals, area chair/senior PC member for conferences, including ACM Multimedia 2012 and ACML'11–'15, technical PC member for many international conferences, and referee for top journals and magazines. He has often been invited for external grant review by worldwide funding agencies, including the US NSF funding agency, Hong Kong RGC funding agency, and so on. He is a senior member of IEEE and a member of AAAI and ACM.

Part I

Introduction

Chapter 1

Introduction

Wall Street is notorious for not learning from its mistakes.
Maybe machines can do better.
– Letting the Machines Decide

Investments in financial markets comprise fundamental and challenging tasks in both financial academia and industry. For example, mutual funds invest the raised capital among a collection of investment opportunities so as to create value for the fund investors; insurance companies invest the premiums among the financial market so as to satisfy the insurance claims in future. Typically, investors analyze and explore investment opportunities via fundamental and technical analyses using various instruments and tools, often done in manual ways. To meet the rapid development of investment opportunities (cf., three challenges in Section 1.1), quantitative analysis has been emerging as a new way for investment analysis and automated trading.

Computational finance (CF), which leverages financial theory via computational techniques, has been emerging and evolving rapidly in recent years. One of the heavily studied areas in CF is the investment, as the computer helps to automate various tasks and make decision in investments. For example, by using advanced computational tools, investment analysts can analyze huge amount of data and identify the under-priced stocks. Besides, investment strategists can back-test and compare strategies using historical data so that they have confidence of the strategy in an unknown future.

One crucial investment task is the allocation of capital, or so-called "portfolio selection" (Markowitz 1952). Despite the theoretical perfectness, estimation errors in their models have constrained their application in real investment. According to DeMiguel et al. (2009), a naive $\frac{1}{N}$ strategies can outperform various portfolio selection models. Such estimation errors lead to portfolio strategies without estimation, or the online portfolio selection (OLPS) pioneered by Cover (1991). We further follow this approach and study the problem of OLPS.

This chapter first introduces the background of OLPS and briefly outlines the contents to be covered in this book.

1.1   Background

The financial investment management industry often faces various challenges and requires new solutions for the task. Below we introduce three representative challenges and briefly propose how to tackle these challenges using machine-learning techniques.

1.1.1   Challenge 1: Voluminous Financial Instruments

One recent challenge is the increasing number of financial instruments,* in terms of both categories and assets in each category. On the one hand, financial innovations (Miller 1986) in the past decade created various types of financial instruments, such as interest rate swaps, credit default swaps, and options. On the other hand, with the development of global economy, thousands of companies and trading instruments are listed on various exchanges.† The "big data" generated by these instruments and companies make it very hard for human investors to process and analyze.

1.1.2   Challenge 2: Human Behavioral Biases

The second challenge is humans behavioral biases in decision making (Barberis and Thaler 2003). Due to humans' subjective nature, many traditional investment strategies suffer from these biases and release sub-optimal decisions when greed and fear interact. Actually, exploiting consistent biases in markets is one source of profits for many traders (Reinganum 1983; Dimson 1988; Jegadeesh 1990). Thus, for an individual investor or institution, it is better to avoid such behavioral biases or even exploit other biases, which is hard for most human investors.

1.1.3   Challenge 3: High-Frequency Trading

The development of information technology has significantly speed up the trading industry. One example is the high-frequency trading (HFT) (Aldridge 2010), which completes the buy and sell within a time ranging from seconds to one day. On the one hand, intraday data are much more voluminous and fast than low-frequency data and, thus, require high-speed tools and methodologies. On the other hand, due to the high speed, HFT requires a quick response to the market behaviors, otherwise the opportunities will disappear. While sometimes human investors can spot the opportunities, they are too slow to open trade positions. Both characteristics of HFT call for new tools and methodologies.

1.1.4   Algorithmic Trading and Machine Learning

To tackle the above three challenges, *algorithmic trading* techniques, which assist investment activities via computational techniques, have emerged. However, with

---

*Financial instruments refer to any tradable assets, such as stocks, futures, and bonds.

†Exchanges provide services, such as trading financial instruments, for traders and brokers. For example, New York Stock Exchange (NYSE) is a stock exchange.

the advance of computational techniques, nowadays machines can handle a much larger quantity of instruments and companies than humans do. It also processes the data in a much higher speed than humans do and is thus suited for HFT scenarios. On the other hand, the machine is free from human behavioral biases and produces exactly the same results if the inputs are the same. There are mainly two areas of algorithmic trading (Harris 2003), one is on the sell side* and the other is on the buy side.[†] The sell side algorithmic trading (Bertsimas and Lo 1998; Almgren and Chriss 2000; Nevmyvaka et al. 2006; Bayraktar 2011) concerns automatically slicing a large order to smaller ones, such that the market impacts incurred by the large order are minimized, while the buy side algorithmic trading (Qian et al. 2007; Chan 2008; Durbin 2010; Kearns et al. 2010) makes intelligent investment decisions to achieve certain targets, such as profit maximization, risk minimization, or both.

Machine learning (Mitchell 1997), a scientific discipline of designing algorithms that can identify complex relationships among huge amounts of historical data and make intelligent decisions upon new data, has been successfully applied to a variety of areas (Manning and Schütze 1999; Baldi and Brunak 2001), including algorithmic trading in finance. For the sell side, there are several patterns among the submitted orders (Harris 2003). To optimally execute one client's large order, machine-learning techniques (Nevmyvaka et al. 2006; Agarwal et al. 2010; Ganchev et al. 2010) can take advantage of the patterns and submit smaller time/volume weighted orders to exchanges, such that the market impacts are minimized. For the buy side, several patterns in financial markets (or, in jargon, "market anomalies") (Dimson 1988; Cont 2001), such as calendar anomalies (Haugen and Lakonishok 1987),[‡] fundamental anomalies (Fama and French 1992),[§] and technical anomalies (Bondt and Thaler 1985; Chan et al. 1996),[¶] are well documented. To generate profits from these patterns, several machine-learning algorithms (El-Yaniv 1998; Yan and Ling 2007; Györfi et al. 2012) have been proposed for buy-side algorithmic trading. Their basic idea is to identify the patterns via machine-learning techniques and obtain profit by trading the patterns.

1.2 What Is Online Portfolio Selection?

This book studies a core problem in the buy-side algorithmic trading named "Online Portfolio Selection" (Cover 1991; Ordentlich and Cover 1996), which sequentially allocates capital among a set of assets aiming to maximize the final return of investment in the long run. OLPS plays a crucial role in a wide range of financial investment

---

*Sell side often refers to investment banks that sell investment services, such as routing orders to exchanges, to asset management firms.

[†]Buy side usually refers to the asset management firms that buy the services from the sell side. For example, Citadel, an asset management firm (buy side), may send their purchase orders via Goldman Sachs, an investment bank (sell side).

[‡]Calendar anomalies refer to the patterns in asset returns from year to year, or month to month. One famous example is the January effect.

[§]Fundamental anomalies are the patterns in asset returns related to the fundamental values of a company, such as size effect and value effect.

[¶]Technical anomalies are patterns related to historical prices, such as momentum, and contrarian.

applications, such as automated wealth management, hedge fund management, and quantitative trading. In the following, to better understand the idea, we begin by introducing a concrete example of real-life OLPS applications.

Suppose Bin, a 30-year-old man, has a capital of $10,000, and he wants to increase the capital to $1,000,000* when he retires at 60 years old, such that he can maintain his current living standards. Assume he has no extra income for investment and purely relies on the initial capital. He would like to achieve this target via the investments in financial markets. Assume that his investment consists of three assets, including Microsoft (stock, symbol: "MSFT"), Goldman Sachs (stock, symbol: "GS"), and Treasury bill.[†] All historical records on the three assets, mainly price quotes, are publicly available. Then, every month,[‡] Bin receives updated information about the three assets and has to face a crucial challenge of decision making, that is, "How to allocate (rebalance) his capital[§] among the three assets every month such that his capital will be more likely increased in the future?" The idea of exploring OLPS technology is to help Bin automate the sequences of allocation/rebalancing decisions so as to maximize his investment return in the long run.

In literature, there are two major schools of principles and theories for portfolio selection: (i) Markowitz's mean variance theory (Markowitz 1952, 1959) that trades off between the expected return (mean) and risk (variance) of a portfolio, which is suitable for single-period portfolio selection and (ii) capital growth theory (or Kelly investment) (Kelly 1956; Breiman 1961; Thorp 1971; Finkelstein and Whitley 1981) that aims to maximize the expected log return of a portfolio and naturally addresses multiple-period investment. Due to the sequential nature of a real-world portfolio selection task, many recent OLPS techniques often design algorithms by following the second family of principles and theories.

Note that this book is focused on the algorithmic aspects, rather than the theory (Breiman 1960; Thorp 1969, 1997; Hakansson 1970, 1971; MacLean et al. 2011). Our study is often concerned with investment management involving multiple types of assets, which may include fixed income securities, equities, and derivatives. Our study is also different from another great body of existing work, which attempted to forecast financial time series by applying computational intelligence techniques and conduct single-stock trading (Katz and McCormick 2000; Huang et al. 2011), such as reinforcement learning (Moody et al. 1998; Moody and Saffell 2001), online prediction (Koolen and Vovk 2012), boosting and expert weighting (Creamer 2007, 2012; Creamer and Freund 2007, 2010; Creamer and Stolfo 2009), neural networks (Kimoto et al. 1993; Dempster et al. 2001), decision trees (Tsang et al. 2004), and support vector machines (Tay and Cao 2001; Cao and Tay 2003; Lu et al. 2009). Finally, we emphasize the nature of "online" algorithms for addressing the portfolio selection problem, in which the algorithms must be computationally efficient enough

---

*Here, one million is an arbitrary number; of course, the more the better.

[†]Treasury bill is often regarded as a risk-free asset, earning a guaranteed risk-free return. Once he does not want to buy any stocks, he can put all money in Treasury bills, instead of cash.

[‡]Here, "month" represents a period, which can be one day, one week, or one month, etc.

[§]For example, he may buy $5000 MSFT stock, $3000 GS stocks, and $2000 Treasury bills.

for handling large-scale applications (e.g., high-frequency trading), although our algorithms are not restricted to high-frequency trading.

1.3 Methodology

OLPS for real-world trading tasks is challenging in that the market information (mainly the market data) arrives sequentially, and a portfolio manager has to make a decision immediately based on the known information. The problem is endogenously online. Two types of machine-learning methodologies have been explored to design strategies for this task.

The first methodology is **batch learning**, where the model is trained from a batch of training instances. In this way, we assume that all price information (and maybe other information) is complete at one decision point, and thus one can deploy batch-learning methods to learn the portfolios. In this mode, one decision is always irrelevant to previous decisions. In particular, we adopt such a mode in one proposed algorithm, which deploys nonparametric learning (or instance-based learning, or case-based learning; Aha 1991; Aha et al. 1991; Cherkassky and Mulier 1998). With an effective trading principle, such a mode can achieve the goal of our project.

The second methodology is **online learning** (or incremental learning), where the model is trained from a single instance in a sequential manner (Shalev-Shwartz 2012; Loveless et al. 2013). Online learning is the process of solving a sequence of problems, given (maybe partial) the solutions to previous problems and possibly additional side information. This definition naturally fits our problem, which is innately online. Contrary to the batch mode, in this mode, one decision is often connected to previous decisions. In particular, in the remaining three of the four algorithms, we adopt two types of online learning techniques (Crammer et al. 2006, 2008, 2009; Dredze et al. 2008) to solve the problem. Besides, to achieve the target of our project, it is also important to exploit an effective trading principle when designing a specific strategy. In this book, we will introduce a variety of classical and modern trading principles that are commonly used for designing OLPS strategies.

After designing a trading strategy, we need to evaluate the effectiveness of the proposed strategy using a back-test methodology. In particular, we feed the historical market data into the testbed to evaluate the strategy and examine how it performs. Through an extensive set of evaluation and analysis of the back-testing performance, we can decide how likely the proposed trading strategy may survive in real-life applications. In this book, we developed an open-source back-testing system, named Online Portfolio Selection, which allows us to benchmark empirical performances of different strategies and algorithms on the same platform. Throughout the book, all the algorithms and strategies will be evaluated on this platform.

1.4 Book Overview

This book consists of four parts, including introduction, principles, algorithms, and empirical studies. Figure 1.1 gives an overview of the book organization of different parts and chapters. The major contents to be covered in each part and each chapter are given below.

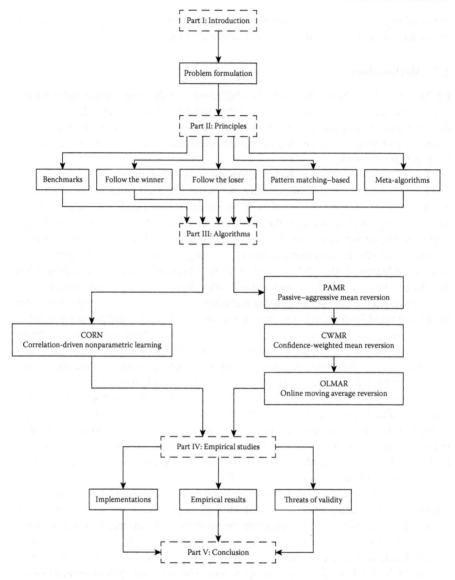

Figure 1.1 *Book organization.*

Part I introduces the background, motivations, and basic definitions of the OLPS problem. Specifically, Chapter 1 introduces the background of computational finance, algorithmic trading, and machine learning and their connections to OLPS. Chapter 2 formally formulates the problem of OLPS as a scientific task.

Part II summarizes the main principles and algorithms of OLPS. In particular, Chapter 3 introduces a family of strategies commonly known as the benchmark principles for OLPS. Chapter 4 introduces the principle of "follow the winner," which is commonly known as the strategies of exploring the "trend following" assumption for investment. Chapter 5 introduces the principles of "follow the loser," which is often known as the strategies of exploring the "mean reversion" assumption for investment. Chapter 6 introduces the principle of pattern matching for OLPS. Finally, Chapter 7 introduces the principle of meta-learning, which attempts to explore the combination of multiple principles and strategies for OLPS.

Part III proposes four OLPS algorithms belonging to two categories, that is, the pattern matching–based approach and follow the loser approach. The first algorithm is a pattern-matching algorithm, "CORrelation-driven Nonparametric learning" (CORN), in Chapter 8. The other three algorithms are mean reversion algorithms. That is, we propose the "passive–aggressive mean reversion" (PAMR) algorithm in Chapter 9, the "confidence-weighted mean reversion" (CWMR) algorithm in Chapter 10, and the "online moving average reversion" (OLMAR) in Chapter 11.

Part IV presents our empirical studies. Chapter 12 introduces the method of empirical studies, and Chapter 13 extensively evaluates the proposed algorithms on real datasets and compares with a set of existing algorithms. Chapter 14 defends the methodologies used in the model setting and empirical studies. Finally, Chapter 15 concludes the book with some future directions.

Chapter 2

Problem Formulation

This chapter introduces the problem setting of online portfolio selection (OLPS) and formally formulates the problem mathematically as a sequential decision task. We further relax the problem setting by adding two practical constraints: transaction costs and margin buying. Finally, we introduce the idea of how to evaluate a strategy's performance.

Specifically, this chapter is organized as follows. Section 2.1 formally formulates the OLPS task as a sequential decision problem. Section 2.2 relaxes the transaction costs and margin buying constraints. Section 2.3 introduces several evaluation metrics for the task. Finally, Section 2.4 summarizes this chapter.

2.1   Problem Settings

Consider an OLPS task: assume an investor aims to invest his capital on a finite number of $m \geq 2$ investment assets* for a finite number of $n \geq 1$ trading periods.[†]

At the t-th period ($t = 1, \ldots, n$), the asset (close) prices are represented by a vector $\mathbf{p}_t \in \mathbb{R}_+^m$, and each element $p_{t,i}$, $i = 1, \ldots, m$, represents the close price of asset i. Their price changes are represented by a *price relative vector* $\mathbf{x}_t \in \mathbb{R}_+^m$, each component of which denotes the ratio of the t-th close price to the last close price, that is, $x_{t,i} = \frac{p_{t,i}}{p_{t-1,i}}$.[‡] Thus, an investment in asset i throughout period t increases by a factor of $x_{t,i}$.[§] Let us denote $\mathbf{x}_1^n = \{\mathbf{x}_1, \ldots, \mathbf{x}_n\}$ as a sequence of price relative vectors for n periods, and $\mathbf{x}_s^e = \{\mathbf{x}_s, \ldots, \mathbf{x}_e\}$, $1 \leq s < e \leq n$ as a market window of price relative vectors ranging from period s to period e.

An investment in the market for the t-th period is specified by a *portfolio vector* $\mathbf{b}_t = (b_{t,1}, \ldots, b_{t,m})$, where $b_{t,i}$, $i = 1, \ldots, m$, represents the proportion of wealth invested in asset i at the beginning of the t-th period. Typically, the portfolio is

---

*When $m = 1$, the problem is reduced to single-stock trading, which is out of the scope of this book.

[†]A period can be a week, a day, an hour, or even a second in high-frequency trading.

[‡]Here we adopt simple gross return, while one may choose simple net return, i.e., $\frac{p_{t,i} - p_{t-1,i}}{p_{t-1,i}}$. For the calculation of the first period, suppose we have $p_{0,i}$.

[§]For example, $x_{t,i} = 2$ means that the investment on an asset will increase by 100%, or double its initial investment. $x_{t,i} = 1$ means that the capital will remain its initial capital.

11

self-financed and no margin/short is allowed; therefore, each entry of a portfolio is nonnegative and adds up to one, that is, $\mathbf{b}_t \in \Delta_m$, where $\Delta_m = \{\mathbf{b}_t : \mathbf{b}_t \succeq 0, \sum_{i=1}^{m} b_{t,i} = 1\}$.* The investment procedure is represented by a *portfolio strategy*, that is, $\mathbf{b}_1 = (\frac{1}{m}, \ldots, \frac{1}{m})$, and the following sequence of mappings:

$$\mathbf{b}_t : \mathbb{R}_+^{m(t-1)} \to \Delta_m, t = 2, 3, \ldots,$$

where $\mathbf{b}_t = \mathbf{b}_t(\mathbf{x}_1^{t-1})$ is the portfolio determined at the beginning of the t-th period upon observing past market behaviors. We denote by $\mathbf{b}_1^n = \{\mathbf{b}_1, \ldots, \mathbf{b}_n\}$ the strategy for n periods, which is the output of an OLPS strategy.

At the t-th period, a portfolio $\mathbf{b}_t$ produces a *portfolio period return* s_t, that is, the wealth increases by a factor of $s_t = \mathbf{b}_t^\top \mathbf{x}_t = \sum_{i=1}^{m} b_{t,i} x_{t,i}$.† Since we reinvest and adopt relative prices, the wealth would grow multiplicatively. Thus, after n periods, a portfolio strategy $\mathbf{b}_1^n$ will produce a *portfolio cumulative wealth* of S_n, which increases the initial wealth by a factor of $\prod_{t=1}^{n} \mathbf{b}_t^\top \mathbf{x}_t$, that is,

$$S_n(\mathbf{b}_1^n, \mathbf{x}_1^n) = S_0 \prod_{t=1}^{n} \mathbf{b}_t^\top \mathbf{x}_t,$$

where S_0 denotes initial wealth and is usually set to $1 for convenience.

We present the framework of the above task in Protocol 2.1. In this task, a portfolio manager's goal is to produce a portfolio strategy ($\mathbf{b}_1^n$) upon the market price relatives ($\mathbf{x}_1^n$), aiming to achieve certain targets. The manager computes the portfolios in a sequential manner. For each period t, the manager has access to the sequence of past price relative vectors $\mathbf{x}_1^{t-1}$. The manager computes a new portfolio $\mathbf{b}_t$ for next price relative vector $\mathbf{x}_t$, where the decision criterion varies among different managers. Then traders will rebalance to the new portfolio, via buying and selling the underlying stocks. At the end of a trading day, the market will reveal $\mathbf{x}_t$. The resulting portfolio $\mathbf{b}_t$ is scored based on portfolio period return s_t. This procedure is repeated until the final period, and the portfolio strategy is scored by its portfolio cumulative wealth S_n.

It is important to note that we make several general and common assumptions in the above model:

1. Transaction cost: no explicit or implicit transaction costs‡ exist.

2. Market liquidity: one can buy and sell the required amount, even fractional, at the last close price of any given trading period.

3. Market impact: any portfolio selection strategy shall not influence the market or any other stocks' prices.

---

*$\succeq 0$ denotes that each element of the vector is nonnegative.

†For example, Bin buys MSFT with 50% of his capital ($5000), GS with 30% of his capital ($3000), and a T-bill with the remaining 20% ($2000). If MSFT goes up by a factor of 2, GS goes down by a factor of 0.5, and the T-bill remains 1. Then his capital will increase by a factor of $0.5 \times 2 + 0.3 \times 0.5 + 0.2 \times 1 = 1.35$, or increase by 35%.

‡Explicit costs include commissions, taxes, stamp duties, and fees. Implicit costs include the bid–ask spread, opportunity costs, and slippage costs.

Protocol 2.1: Online portfolio selection.

Input: $\mathbf{x}_1^n$: Historical market price relative sequence
Output: S_n: Final cumulative wealth

Initialize $S_0 = 1, \mathbf{b}_1 = \left(\frac{1}{m}, \ldots, \frac{1}{m} \right)$
for $t = 1, 2, \ldots, n$ **do**
> Portfolio manager learns a portfolio $\mathbf{b}_t$;
> Market reveals a price relative vector $\mathbf{x}_t$;
> Portfolio incurs period return $s_t = \mathbf{b}_t^\top \mathbf{x}_t$ and updates cumulative return
> $S_t = S_{t-1} \times \left(\mathbf{b}_t^\top \mathbf{x}_t \right)$;
> Portfolio manager updates his or her decision rules;

end

The preceding assumptions are nontrivial. We will further analyze and discuss their implications and effects for our empirical studies in Sections 13.4 and 14.1.

Finally, as we are going to design intelligent learning algorithms that fit the above model, let us fix the objective of the proposed learning algorithms. For a portfolio selection task, one can choose to maximize risk-adjusted return (Markowitz 1952; Sharpe 1964) or to maximize cumulative return (Kelly 1956; Thorp 1971) at the end of a period. While the model is online, which contains multiple periods, we choose to maximize the cumulative return (Hakansson 1971),* which is also the objective of most existing algorithmic studies.

2.2   Transaction Costs and Margin Buying Models

While our model is concise and simple to understand, it ignores some practical issues in real trading scenarios. We now relax two constraints to address these issues.

In reality, an important and unavoidable issue is the *transaction cost*, which includes the commission fees and taxes imposed by brokers and governments, during the rebalance activities.[†] Note that the transaction cost is imposed by markets, and a portfolio's behavior cannot change the properties of transaction costs, such as commission rates or tax rates. To handle the issue, the first way, which is commonly adopted by existing strategies, is that a portfolio selection model does not take transaction costs into account, and the second way is to directly integrate the costs in the model (Györfi and Vajda 2008). In this book, we take the first way and adopt the simplified *proportional transaction cost* model (Blum and Kalai 1999; Borodin et al. 2004).[‡] To be specific, rebalancing a portfolio incurs transaction costs

---

*However, such objective does not prohibit us from comparing different strategies via risk-adjusted terms, such as the Sharpe ratio and Calmar ratio.

†Besides commission and taxes, some other factors, such as bid–ask spreads, also implicitly incur transaction costs to a portfolio.

‡Blum and Kalai (1999, p. 195) also provide the precise model, with which our algorithms' performance is almost the same as the simplified one, but the precise model costs much more time to compute.

on every buy and sell operation, based upon a transaction cost rate of $\gamma \in (0, 1)$. At the beginning of period t, the portfolio manager rebalances his or her wealth to a new portfolio $\mathbf{b}_t$, from last close price adjusted portfolio $\hat{\mathbf{b}}_{t-1}$, each component of which is calculated as $\hat{b}_{t-1,i} = \frac{b_{t-1,i} \times x_{t-1,i}}{\mathbf{b}_{t-1}^\top \mathbf{x}_{t-1}}$. Such rebalance incurs a transaction cost of $\frac{\gamma}{2} \times \sum_{i=1}^{m} |b_{t,i} - \hat{b}_{t-1,i}|$, where the initial portfolio is set to $(0, \ldots, 0)$. Thus, the cumulative wealth after n periods can be expressed as

$$S_n^\gamma = S_0 \prod_{t=1}^{n} \left[(\mathbf{b}_t \cdot \mathbf{x}_t) \times \left(1 - \frac{\gamma}{2} \times \sum_{i=1}^{m} |b_{t,i} - \hat{b}_{t-1,i}| \right) \right].$$

Another practical issue is *margin buying*, which allows the portfolio managers to buy securities with cash borrowed from securities brokers, using their own equity positions as collateral. Following existing studies (Cover 1991; Helmbold et al. 1998; Agarwal et al. 2006), we relax this constraint and evaluate it empirically. We assume the margin setting to be 50% down and a 50% loan,* at an annual interest rate of 6% (equivalently, the corresponding daily interest rate of borrowing, c, is set to 0.000238). With such a setting, a new asset named "margin component" is generated for each asset, and its price relative for period t equals $2 \times x_{t,i} - 1 - c$. In the case of $x_{t,i} \leq \frac{1+c}{2}$, which means the stock drops more than half, we simple set its margin component to 0 (Li et al. 2012).[†] As a result, if margin buying is allowed, the total number of assets becomes $2m$. By adding such a "margin component," we can magnify both the potential profit or loss on the i-th asset.[‡]

2.3 Evaluation

One standard criterion to evaluate an OLPS strategy is its *portfolio cumulative wealth* at the end of trading periods. As we set the initial wealth, $S_0 = 1$ and thus S_n also denote the *portfolio cumulative return*, which is the ratio of final portfolio cumulative wealth divided by its initial wealth. Another equivalent criterion, which considers compounding effect, is *annualized percentage yield* (APY), that is, $\text{APY} = \sqrt[y]{S_n} - 1$, where y is the number of years corresponding to n periods.[§] APY measures the average wealth increment that a strategy could achieve in a year. Typically, the higher the *portfolio cumulative wealth* or *annualized percentage yield*, the better the strategy's performance is.

Besides the absolute return metrics, it is also important to evaluate a strategy's *risk* and *risk-adjusted return* (Sharpe 1963, 1994). One common criterion is the *annualized standard deviation* of portfolio period returns to measure volatility risk and

---

*That is, if one has \$100 stock (down or collateral) one can borrow at most \$100 cash (loan).

[†] Such a measure is not perfect since it manually changes the margin component, although less than 5 per dataset. One may refer to Györfi et al. (2012, Chapter 4) for other solutions to the possibility of ruin.

[‡] For example, assume two assets with price relatives of $(1.1, 0.9)$. After adjustment, the price relative vector becomes $(1.1, 0.9, 1.2, 0.8)$. Putting wealth on the latter two margin components, the portfolio's profit or loss magnifies. That is, 10% profit (1.1) becomes 20% $(1.1 \times 2 - 1 - c)$ and 10% loss (0.9) also becomes 20% $(0.9 \times 2 - 1 - c)$. Note that the portfolio vector representing the proportions of capital is still a simplex.

[§] One year consists of 252 trading days or 50 trading weeks.

the *annualized Sharpe ratio* (SR) (Sharpe 1966) to evaluate volatility risk-adjusted return. To obtain the *annualized standard deviation*, we calculate the standard deviation of daily returns and multiply by $\sqrt{252}$.* For volatility risk-adjusted return, we calculate the *annualized* SR as

$$SR = \frac{APY - R_f}{\sigma_p},$$

where R_f is the risk-free return[†] and σ_p is the annualized standard deviation. The higher the *annualized* SR, the better the strategy's (volatility) risk-adjusted return is.

Portfolio management community often conducts drawdown analysis (Magdon-Ismail and Atiya 2004) to measure the decline from a historical peak of portfolio cumulative wealth. Formally, a strategy's *drawdown* (DD) at period t is defined as $DD(t) = \sup[0, \sup_{i \in (0,t)} S_i - S_t]$. Its *maximum drawdown* (MDD) is the maximum of drawdowns over all periods and can effectively measure a strategy's downside risk. Formally, *maximum drawdown* for a horizon of n, $MDD(n)$, is defined as

$$MDD(n) = \sup_{t \in (0,n)} [DD(t)].$$

Moreover, practitioners also adopt the *Calmar ratio* (CR) (Young 1991) to measure a strategy's drawdown risk-adjusted return:

$$CR = \frac{APY}{MDD}.$$

The smaller the *maximum drawdown*, the more drawdown risk the strategy can tolerate. The higher the *Calmar ratio*, the better (drawdown) risk-adjusted return the strategy is.

To test whether simple luck can generate the return achieved by a strategy, portfolio management practitioners (Grinold and Kahn 1999) can conduct statistical tests. Since all test datasets are just samples of the market population, such tests can validate a strategy for future. We conduct a *Student's t-test* to determine the likelihood that the observed profitability is due to chance alone (under the assumption that a strategy is not profitable in the population). Since the sample profitability is being compared with no profitability, 0 is subtracted from the sample mean profit/loss. Note that (daily) profit/loss equals (daily) return minus 1. The standard error of mean is calculated as the standard deviation divided by square root of the number of periods. The t-statistic is the sample profit mean[‡] divided by the sample standard error. Finally, the probability of the t-statistic can be calculated with a degree of freedom equal to the number of periods minus 1. Note that the *Student's t-test* assumes that the underlying distribution of data is normal. According to the central limit theorem, as the sample size increases, the distribution of the sample mean approaches normal. If a sample

---

*Here, 252 denotes the average number of annual trading days. For other frequencies, we can choose their corresponding numbers.

[†]Typically, it equals the return of Treasury bills, and we fix it at 4% per year, or 0.000159 per day.

[‡]Suppose we compare the sample profit mean with 0.

dataset contains a large number of trading transactions, which is often the case in our empirical evaluations, we could regard the distribution of profit/loss as normal. The smaller the probability, the higher the confidence we have toward the strategy.

2.4 Summary

Online portfolio selection (OLPS) is a fundamental and practical computational finance problem. It can be mathematically formulated as a sequential decision task that aims to decide the best sequence of decisions to maximize the investment goals in the long run. It has been extensively studied in the literature, and recent years have witnessed a rapid growth of fruitful research achievements. The next part will introduce a family of important principles widely used for solving this challenging task.

Part II

Principles

Existing online portfolio selection (OLPS) approaches follow the preceding problem formulations and derive explicit portfolio update schemes. Table II.1 (Li and Hoi 2014) summarizes the main principles and several representative algorithms, and four of the algorithms are illustrated in detail in later chapters. In particular, first we introduce several *benchmark* algorithms. Then, we introduce three categories of principles or algorithms with explicit portfolio update schemes, which are classified according to the directions of weight transfer. The first approach, *follow the winner*, increases the weights of more successful experts or stocks, often based on their historical performance. Contrarily, the second approach, *follow the loser*, increases the weights of less successful experts or stocks, or transfers the weights from winners to losers. The third category, the *pattern matching–based approach*, constructs portfolios based on similar historical patterns and has no explicit directions. Finally, we survey some related *meta-algorithms* applying to a set of experts, each of which is equipped with any algorithms in the preceding three categories.

This part is organized as follows. Chapter 3 surveys the benchmarks used in this study. Chapter 4 surveys the first principle, follow the winner. Chapter 5 surveys the second principle, follow the loser. Then, Chapter 6 introduces the third principle, pattern matching–based approaches. The final principle, meta-algorithms, is introduced in Chapter 7.

Table II.1 *Principles and representative online portfolio selection algorithms*

Classifications	Algorithms	Representative References
Benchmarks	Buy and Hold	
	Best Stock	
	Constant Rebalanced Portfolios	Kelly (1956); Cover (1991)
Follow the winner	Universal Portfolios	Cover (1991)
	Exponential Gradient	Helmbold et al. (1998)
	Follow the Leader	Gaivoronski and Stella (2000)
	Follow the Regularized Leader	Agarwal et al. (2006)
	Aggregating-Type Algorithms	Vovk and Watkins (1998)
Follow the loser	Anticorrelation	Borodin et al. (2004)
	Passive–Aggressive Mean Reversion	Li et al. (2012)
	Confidence Weighted Mean Reversion	Li et al. (2013)
	Online Moving Average Reversion	Li et al. (2015)
	Robust Median Reversion	Huang et al. (2013)

(Continued)

Table II.1 *(Continued) Principles and representative online portfolio selection algorithms*

Classifications	Algorithms	Representative References
Pattern matching–based approaches	Nonparametric Histogram Log-Optimal Strategy	Györfi et al. (2006)
	Nonparametric Kernel-Based Log-Optimal Strategy	Györfi et al. (2006)
	Nonparametric Nearest Neighbor Log-Optimal Strategy	Györfi et al. (2008)
	Correlation-Driven Nonparametric Learning Strategy	Li et al. (2011a)
	Nonparametric Kernel-Based Semi-Log-Optimal Strategy	Györfi et al. (2007)
	Nonparametric Kernel-Based Markowitz-Type Strategy	Ottucsák and Vajda (2007)
	Nonparametric Kernel-Based GV-Type Strategy	Györfi and Vajda (2008)
Meta-algorithms	Aggregating Algorithm	Vovk (1990); Vovk and Watkins (1998)
	Fast Universalization Algorithm	Akcoglu et al. (2005)
	Online Gradient Updates	Das and Banerjee (2011)
	Online Newton Updates	Das and Banerjee (2011)
	Follow the Leading History	Hazan and Seshadhri (2009)

Source: Li and Hoi (2014).

Chapter 3

Benchmarks

3.1 Buy-and-Hold Strategy

The most common baseline is the *Buy-and-Hold* (BAH) strategy, in which one invests wealth among the market with an initial portfolio of $\mathbf{b}_1$ and holds the portfolio till the end. The manager only buys the assets at the beginning of the first period and does not rebalance in subsequent periods, while the portfolio holdings are implicitly changed following the market fluctuations. In particular, at the end of period t, the portfolio holding becomes $\frac{\mathbf{b}_t \odot \mathbf{x}_t}{\mathbf{b}_t^\top \mathbf{x}_t}$, where $\odot$ denotes the element-wise product.* BAH's final cumulative wealth is the initial portfolio-weighted average of individual asset returns, that is,

$$S_n(\text{BAH}(\mathbf{b}_1)) = \mathbf{b}_1 \cdot \left(\bigodot_{t=1}^{n} \mathbf{x}_t \right),$$

where $\mathbf{b} \cdot \mathbf{x}$ denotes the inner product $\mathbf{b}^\top \mathbf{x}$. The BAH strategy with uniform portfolio $\mathbf{b}_1 = \left(\frac{1}{m}, \ldots, \frac{1}{m}\right)$ is referred to as the *uniform BAH* strategy, which is usually adopted as a *market* strategy to produce a market index.[†]

3.2 Best Stock Strategy

Another common benchmark is the *Best Stock* (Best) strategy, which is a special BAH strategy that invests all capital on the best stock in hindsight. Its initial portfolio $\mathbf{b}^0$ can be calculated as, $\mathbf{b}^0 = \arg\max_{\mathbf{b} \in \Delta_m} \mathbf{b} \cdot \left(\odot_{t=1}^{n} \mathbf{x}_t \right)$, which is thus a hindsight strategy. The strategy's final cumulative wealth equals

$$S_n(\text{Best}) = \max_{\mathbf{b} \in \Delta_m} \mathbf{b} \cdot \left(\bigodot_{t=1}^{n} \mathbf{x}_t \right) = S_n(\text{BAH}(\mathbf{b}^0)).$$

---

*For example, assuming two assets with price relative vectors $(2, 1)$ and the portfolio at the beginning of a period is $(0.5, 0.5)$, then the actual weights at the end of the period becomes $\frac{(0.5 \times 2, 0.5 \times 1)}{0.5 \times 2 + 0.5 \times 1} = (0.67, 0.33)$.

†Market index can also be calculated using other methods, such as capitalization weighted index and market share weighted index.

3.3 Constant Rebalanced Portfolios

One challenging benchmark is the *Constant Rebalanced Portfolios* (CRP) strategy, which rebalances to a fixed portfolio **b** every period.* In particular, the portfolio strategy can be represented as $\mathbf{b}_1^n = \{\mathbf{b}, \mathbf{b}, \ldots, \mathbf{b}\}$, in which **b** is a predefined portfolio. Thus, CRP's final cumulative portfolio wealth can be calculated as

$$S_n(\text{CRP}(\mathbf{b})) = \prod_{t=1}^{n} \mathbf{b}^\top \mathbf{x}_t.$$

One special CRP with uniform portfolio $\mathbf{b} = \left(\frac{1}{m}, \ldots, \frac{1}{m}\right)$ is named as *Uniform Constant Rebalanced Portfolios* (UCRP). Another special CRP is the optimal offline[†] CRP strategy, whose portfolio can be calculated as

$$\mathbf{b}^\star = \arg\max_{\mathbf{b}^n \in \Delta_m} S_n(\text{CRP}(\mathbf{b})) = \arg\max_{\mathbf{b} \in \Delta_m} \prod_{t=1}^{n} (\mathbf{b}^\top \mathbf{x}_t),$$

which is convex and can be efficiently solved. The CRP with $\mathbf{b}^\star$ is denoted as *Best Constant Rebalanced Portfolios* (BCRPs), which achieve a final cumulative wealth as

$$S_n(\text{BCRP}) = \max_{\mathbf{b} \in \Delta_m} S_n(\text{CRP}(\mathbf{b})) = S_n(\text{CRP}(\mathbf{b}^\star))$$

Note that BCRP is a hindsight strategy, which can only be calculated with complete market sequences. Cover (1991) proved that BCRP is the best strategy in an independent and identically distributed (i.i.d.) market and showed its benefits as a target, that is, BCRP exceeds the Best Stock strategy, Value Line Index (geometric mean of asset returns), and Dow Jones Index (arithmetic mean of asset returns, or BAH). In addition, BCRP is invariant under permutations of market sequence, that is, it does not depend on the order in which $\mathbf{x}_1, \mathbf{x}_2, \ldots, \mathbf{x}_n$ occur.

One desired theoretical result for an OLPS algorithm is *universality* (Cover 1991; Ordentlich 2010). An algorithm *Alg* is *universal* if the average *(external) regret* (Stoltz and Lugosi 2005; Blum and Mansour 2007) for n periods asymptotically approaches 0, that is,

$$\frac{1}{n}\text{regret}_n(Alg) = \frac{1}{n}(\log S_n(\text{BCRP}) - \log S_n(Alg)) \xrightarrow{n \to \infty} 0. \qquad (3.1)$$

In other words, for an arbitrary sequence of price relatives, a universal algorithm asymptotically approaches the same exponential growth rate as the BCRP strategy.

Since CRP rebalances to a fixed portfolio each period, its frequent transactions will incur high transaction costs. Helmbold et al. (1998) proposed a *Semi-Constant Rebalanced Portfolio*, which rebalances on selected periods rather than every period.

---

*CRP differs from BAH as the former actively rebalances to a predefined portfolio for every period, while the latter does not rebalance during the entire trading period. However, the portfolio holding of BAH passively changes as the stock prices fluctuate.

[†]Contrary to the online case, offline assumes that all price relatives over the n periods are available.

Chapter 4

Follow the Winner

The first principle, *follow the winner*, is characterized by increasing the weights of more successful experts or stocks. Rather than targeting market and best stock, algorithms in this category often aim to track the BCRP strategy, that is, their target is to be *universal*.

This chapter is organized as follows. Section 4.1 introduces Cover's universal portfolios (UP) algorithm, and Section 4.2 details the exponential gradient (EG) algorithm. Sections 4.3 and 4.4 introduce the follow the leader (FTL) and follow the regularized leader (FTRL) approaches, respectively. Finally, Section 4.5 summarizes the follow the winner principle.

4.1   Universal Portfolios

The basic idea of *Universal Portfolios*–type (UP-type) algorithms is to assign the capital to base experts of a single class, let the experts run, and finally pool their wealth. They are analogous to the *Buy-and-Hold* (BAH) strategy. In particular, BAH's base experts belong to a special strategy investing on a single asset, and thus the number of experts equals that of assets. In other words, BAH buys individual stocks, lets the stocks go, and finally pools their individual wealth. On the other hand, the base experts in the UP-type algorithms can be any single strategy class that invests over the whole markets. Besides, UP-type algorithms are also similar to the *meta-algorithms* (MA) in Chapter 7, while the latter applies to base experts of multiple classes.

Cover (1991) proposed the *universal portfolio* strategy, and Cover and Ordentlich (1996) further refined the algorithm as μ-*weighted universal portfolio*, in which μ denotes a given distribution on the space of valid portfolio Δ_m. Intuitively, Cover's UP operates similar to a *fund of funds* (FOF),* and its main idea is to buy and hold parameterized CRP strategies over the whole simplex domain. In particular, it initially invests a proportion of wealth $d\mu(\mathbf{b})$ to each portfolio manager operating CRP strategy with $\mathbf{b} \in \Delta_m$, and lets the CRP managers run. Then, at the end, each manager will grow his wealth to $S_n(\mathbf{b})d\mu(\mathbf{b})$. Finally, Cover's UP pools the individual experts'

---

*An FOF holds a portfolio of other investment funds, rather than directly investing in stocks, futures, etc.

wealth over the continuum of portfolio strategies. Note that $S_n(\mathbf{b}) = e^{n W_n(\mathbf{b})}$, which means that the portfolio grows at an exponential rate of $W_n(\mathbf{b})$.

Formally, its update scheme (Cover and Ordentlich 1996, Definition 1) can be interpreted as a historical performance-weighted average of all valid constant rebalanced portfolios,

$$\mathbf{b}_{t+1} = \frac{\int_{\Delta_m} \mathbf{b} S_t(\mathbf{b}) d\mu(\mathbf{b})}{\int_{\Delta_m} S_t(\mathbf{b}) d\mu(\mathbf{b})}.$$

Note that at the beginning of period $t+1$, one CRP manager's wealth (historical performance) equals $S_t(\mathbf{b})d\mu(\mathbf{b})$. Incorporating the initial wealth of $S_0 = 1$, the final cumulative wealth is the weighted average of CRP managers' wealth (Cover and Ordentlich 1996, Eq. (24)):

$$S_n(UP) = \int_{\Delta_m} S_n(\mathbf{b}) d\mu(\mathbf{b}). \tag{4.1}$$

One special case is that μ equals a uniform distribution; the portfolio update reduces to Cover's UP (Cover 1991, Eq. (1.3)). Another special cases is Dirichlet $\left(\frac{1}{2}, \ldots, \frac{1}{2}\right)$ weighted UP (Cover and Ordentlich 1996), which is proved to be a more optimal allocation.

Alternatively, if a loss function is defined as the negative logarithmic function of portfolio return, Cover's UP is actually an exponentially weighted average forecaster (Cesa-Bianchi and Lugosi 2006). The regret (Cover 1991) achieved by Cover's UP is $O(m \log n)$, and its time complexity is $O(n^m)$, where m denotes the number of stocks and n refers to the number of periods. Cover and Ordentlich (1996) proved that the $\left(\frac{1}{2}, \ldots, \frac{1}{2}\right)$ weighted UP has the same scale of regret bound, but a better constant term (Cover and Ordentlich 1996, Theorem 2).

As Cover's UP is based on an ideal market model, one research direction is to extend the algorithm to handle various realistic assumptions. Cover and Ordentlich (1996) considered side information, including experts' opinions and fundamental data. Cover and Ordentlich (1998) extended the algorithm to handle short selling and margin, and Blum and Kalai (1999) took account of transaction costs.

Another research direction is to generalize Cover's UP with different base classes, rather than the CRP strategy. Jamshidian (1992) generalized the algorithm for continuous time markets and presented its long-term performance. Vovk and Watkins (1998) applied the *aggregating algorithm* (AA) (Vovk 1990) to a finite number of arbitrary investment strategies, of which Cover's UP becomes a specialized case when applied to an infinite number of CRPs. Ordentlich and Cover (1998) analyzed the minimal ratio of final wealth achieved by any nonanticipating investment strategy to that of BCRP and presented a strategy to achieve such an optimal ratio. Cross and Barron (2003) generalized Cover's UP from the CRP strategy class to any parameterized target class and proposed a computation favorable universal strategy. Akcoglu et al. (2005) extended Cover's UP from a parameterized CRP class to a wide class of investment strategies, including trading strategies operating on a single stock and portfolio strategies operating on the whole stock market. Kozat and Singer (2011) proposed a similar universal algorithm based on the class of semiconstant rebalanced portfolios (Helmbold et al. 1998), which provides good asymptotic performance in case of nonzero transaction costs.

Besides our intuitive analysis, various works have been proposed to discuss the connection between Cover's UP with universal prediction (Feder et al. 1992), data compression (Rissanen 1983), and Markowitz's mean variance theory (Markowitz 1952). Algoet (1992) discussed a universal scheme for prediction, gambling, and portfolio selection. Cover (1996) and Ordentlich (1996) discussed the connection between UP selection and data compression. Belentepe (2005) presented a statistical view of Cover's UP strategy and claimed its approximate equivalence to the mean variance portfolio theory.

Although Cover's UP has a tight regret bound, its implementation is exponential to the number of stocks, which restricts its practical applicability. To handle the computational issue, Kalai and Vempala (2002) presented an efficient implementation based on rapidly mixing nonuniform random walks, improving the running time from original $O(n^m)$ to $O(m^7 n^8)$.

4.2   Exponential Gradient

Algorithms in the *exponential gradient* type (EG type) focus on the following optimization formulation:

$$\mathbf{b}_{t+1} = \underset{\mathbf{b} \in \Delta_m}{\arg \max} \, \eta \log \mathbf{b} \cdot \mathbf{x}_t - R(\mathbf{b}, \mathbf{b}_t), \qquad (4.2)$$

where $R(\mathbf{b}, \mathbf{b}_t)$ denotes a regularization term and $\eta > 0$ denotes a learning rate. One straightforward interpretation is to track the best stock in the last period while keeping previous portfolio information via a regularization term.

Helmbold et al. (1998) proposed the EG strategy, which is based on the same algorithm for mixture estimation (Helmbold et al. 1997). Following Equation 4.2, EG adopts relative entropy as its regularization term, that is,

$$R(\mathbf{b}, \mathbf{b}_t) = \sum_{i=1}^{m} b_i \log \frac{b_i}{b_{t,i}}.$$

EG's formulation is convex in $\mathbf{b}$; however, it is hard to solve since the log function is nonlinear. Thus, the authors adopted log's first-order Taylor expansion at $\mathbf{b}_t$, that is,

$$\log \mathbf{b} \cdot \mathbf{x}_t \approx \log(\mathbf{b}_t \cdot \mathbf{x}_t) + \frac{\mathbf{x}_t}{\mathbf{b}_t \cdot \mathbf{x}_t}(\mathbf{b} - \mathbf{b}_t).$$

Then the nonlinear log term becomes linear and the optimization is easy to solve. Solving the optimization, we can obtain EG's update rule as

$$b_{t+1,i} = b_{t,i} \exp\left(\eta \frac{x_{t,i}}{\mathbf{b}_t \cdot \mathbf{x}_t}\right) \Big/ Z, \quad i = 1, \ldots, m,$$

where Z denotes the normalization term such that the portfolio weights sum to 1.

Besides the multiplicative update rule (EG), the optimization problem can also be solved using the *gradient projection* (GP) and *expectation–maximization* (EM) (Helmbold et al. 1997). Rather than EG's relative entropy, GP adopts an L2-norm regularization, and EM adopts an χ^2 regularization term, that is,

$$R(\mathbf{b}, \mathbf{b}_t) = \begin{cases} \frac{1}{2} \sum_{i=1}^{m} (b_i - b_{t,i})^2 & \text{GP} \\ \frac{1}{2} \sum_{i=1}^{m} \frac{(b_i - b_{t,i})^2}{b_{t,i}} & \text{EM} \end{cases}.$$

Solving corresponding optimization problems, we can obtain GP's update rule as

$$b_{t+1,i} = b_{t,i} + \eta \left(\frac{x_{t,i}}{\mathbf{b}_t \cdot \mathbf{x}_t} - \frac{1}{m} \sum_{j=1}^{m} \frac{x_{t,j}}{\mathbf{b}_t \cdot \mathbf{x}_t} \right),$$

and EM's update rule as

$$b_{t+1,i} = b_{t,i} \left(\eta \left(\frac{x_{t,i}}{\mathbf{b}_t \cdot \mathbf{x}_t} - 1 \right) + 1 \right).$$

The latter can also be viewed as EG's first-order approximation.

One key parameter for the EG-type algorithms is the learning rate η. To achieve a universal regret bound, η has to be small. However, as $\eta \to 0$, its update approaches uniform,* which degrades to UCRP. Such an analysis will be empirically verified in Section 13.6.

EG has a regret bound of $O(\sqrt{n \log m})$ and a running time of $O(mn)$. The regret is not as tight as Cover's UP; however, its linear time substantially surpasses that of UP. Besides, the authors also proposed a variant by transforming all price relatives, which has a tight regret bound of $O(m \log n)$. Though not officially proposed for online portfolio selection (Helmbold et al. 1997), GP can straightforwardly achieve a regret of $O(\sqrt{mn})$, which is significantly worse than EG.

Das and Banerjee (2011) generalized EG-type algorithms to an MA named *online gradient updates*, which combine underlying experts such that the overall system performs no worse than any convex combination of its base experts.

4.3   Follow the Leader

FTL strategies directly track the BCRP till time t:

$$\mathbf{b}_{t+1} = \mathbf{b}_t^* = \arg\max_{\mathbf{b} \in \Delta_m} \sum_{\tau=1}^{t} \log(\mathbf{b} \cdot \mathbf{x}_\tau). \tag{4.3}$$

Intuitively, this category follows the BCRP leader over the known periods, and the ultimate leader is BCRP over the whole periods.

---

*In case of $\eta = 0$, $b_{t+1,i} = b_{t,i} = \cdots = b_{1,i} = \frac{1}{m}$.

Ordentlich (1996, Chapter 4.4) briefly mentioned a strategy to obtain portfolios by mixing BCRP to date and uniform portfolio:

$$\mathbf{b}_{t+1} = \frac{t}{t+1}\mathbf{b}_t^* + \frac{1}{t+1}\frac{1}{m}\mathbf{1}.$$

However, its worst-case regret bound is worse than that of Cover's UP.

Gaivoronski and Stella (2000) proposed *successive constant rebalanced portfolios* (SCRPs) and *weighted successive constant rebalanced portfolios* (WSCRPs) for stationary markets. For each period, SCRP directly adopts the BCRP to date, $\mathbf{b}_{t+1} = \mathbf{b}_t^*$. The authors further solved the optimal portfolio $\mathbf{b}_t^*$ via stochastic optimization (Birge and Louveaux 1997), resulting in the updates (Gaivoronski and Stella 2000, Algorithm 1). On the other hand, WSCRP outputs a convex combination of the SCRP and previous portfolio:

$$\mathbf{b}_{t+1} = (1 - \gamma)\mathbf{b}_t^* + \gamma\mathbf{b}_t,$$

where $\gamma \in [0, 1]$ represents a trade-off parameter. The regret bounds achieved by SCRP and WSCRP are $O(m \log n)$, which is the same as that of Cover's UP.

Rather than assuming a stationary market, some algorithms in this category assume that the historical market is nonstationary. Gaivoronski and Stella (2000) proposed *variable rebalanced portfolios* (VRP), which calculate the BCRP on a latest sliding window. To be specific, VRP updates the portfolio as

$$\mathbf{b}_{t+1} = \arg\max_{\mathbf{b}\in\Delta_m} \sum_{\tau=t-W+1}^{t} \log(\mathbf{b} \cdot \mathbf{x}_\tau),$$

where W denotes a specified window size.

Gaivoronski and Stella (2003) further proposed *adaptive portfolio selection* (APS). By changing the objective, APS can handle three portfolio selection tasks, that is, adaptive Markowitz portfolio, log-optimal constant rebalanced portfolio, and index tracking. To handle the transaction cost issue, they further proposed *threshold portfolio selection*, which only rebalances the portfolio if the expected return of a new portfolio exceeds that of the last portfolio by a threshold.

4.4 Follow the Regularized Leader

The FTRL approach adds a regularization term to Equation 4.3:

$$\mathbf{b}_{t+1} = \arg\max_{\mathbf{b}\in\Delta_m} \sum_{\tau=1}^{t} \log(\mathbf{b} \cdot \mathbf{x}_\tau) - \frac{\beta}{2}R(\mathbf{b}), \tag{4.4}$$

where β denotes a trade-off parameter and $R(\mathbf{b})$ is the regularization term on $\mathbf{b}$. Note that the first term includes all historical information; thus, the regularization term only relates to the next portfolio, which is different from the EG algorithm. One typical regularization is L2-norm, that is, $R(\mathbf{b}) = \|\mathbf{b}\|^2$.

Agarwal et al. (2006) proposed the *online Newton step* (ONS), by solving the optimization problem (4.4) with L2-norm regularization via online convex optimization (Zinkevich 2003; Hazan 2006; Hazan et al. 2006, 2007). Similar to the regular offline Newton method, the basic idea of ONS is to replace the log term via its second-order Taylor expansion at $\mathbf{b}_t$, and then solve for the closed-form updates. Finally, the update rule of ONS is

$$\mathbf{b}_1 = \left(\frac{1}{m}, \dots, \frac{1}{m} \right), \quad \mathbf{b}_{t+1} = \Pi_{\Delta_m}^{\mathbf{A}_t} (\delta \mathbf{A}_t^{-1} \mathbf{c}_t),$$

with $\mathbf{A}_t = \sum_{\tau=1}^t \left(\frac{\mathbf{x}_\tau \mathbf{x}_\tau^\top}{(\mathbf{b}_\tau \cdot \mathbf{x}_\tau)^2} \right) + \mathbf{I}_m$ and $\mathbf{c}_t = \left(1 + \frac{1}{\beta} \right) \sum_{\tau=1}^t \frac{\mathbf{x}_\tau}{\mathbf{b}_\tau \cdot \mathbf{x}_\tau}$, where β is a trade-off parameter, δ is a scaling term, $\mathbf{I}_m$ denotes an $m \times m$ diagonal matrix, and $\Pi_{\Delta_m}^{\mathbf{A}_t} (\cdot)$ is an exact projection to the simplex domain.

ONS's regret bound is $O(m^{1.5} \log(mn))$, which is slightly worse than that of Cover's UP. Since it iteratively updates the first- and second-order information, it costs $O(m^3)$ per period, which is irrelevant to the number of periods. To sum up, its total time cost is $O(m^3 n)$.

While FTRL focuses on the worst-case investing, Hazan and Kale (2009, 2012) linked the worst-case investing with practically widely used average-case investing, that is, the geometric Brownian motion (GBM) model (Bachelier 1900; Osborne 1959; Cootner 1964). The authors designed an investment strategy that is universal in the worst case and is capable of exploiting the GBM model. The algorithm, or so-called *Exp-Concave-FTL*, follows a similar formulation to ONS, that is,

$$\mathbf{b}_{t+1} = \arg\max_{\mathbf{b} \in \Delta_m} \sum_{\tau=1}^t \log(\mathbf{b} \cdot \mathbf{x}_\tau) - \frac{1}{2} \|\mathbf{b}\|^2.$$

The optimization problem can be efficiently solved via online convex optimization, which typically requires a high time complexity (i.e., similar to the ONS). If the stock price follows the GBM model, the regret round becomes $O(m \log Q)$, where Q is a quadratic variability calculated as $n - 1$ times the sample variance of price relative vectors. Since Q is typically much smaller than n, the regret bound is significantly improved from previous $O(m \log n)$.

Besides the improved regret bound, the authors also discussed the relationship between their algorithm and trading frequency. The authors asserted that increasing the trading frequency would decrease the variance of minimum-variance CRP, while the regret stays the same. Therefore, it is expected to see improved performance as the trading frequency increases, which is empirically observed by Agarwal et al. (2006).

Das and Banerjee (2011) further extended the FTRL approach to a generalized MA termed *online Newton update* (ONU), which guarantees that the overall performance is no worse than any convex combination of the base experts.

4.5 Summary

Follow the winner is the main principle of online portfolio selection research. While most algorithms in this category are guaranteed by the theory (regret bound), their empirical performance is not outstanding (cf. the empirical results in Chapter 13). We believe that the main reason for this phenomenon is that their target in hindsight, or BCRP, assumes that price relatives follow i.i.d., which may be contradictory to the empirical evidence of real markets. The next chapter will introduce a different principle, follow the loser, which makes a different assumption regarding market behaviors.

Chapter 5

Follow the Loser

The Best Constant Rebalanced Portfolios (BCRP) strategy is optimal if the market is independent and identically distributed (i.i.d.; Cover 1991); however, this assumption may not fit the real market and thus may lead to the inferior performance of the "follow the winner" category. Rather than tracking the winners, the *follow the loser* approach is often characterized by transferring the wealth from winners to losers. The underlying assumption is the *mean reversion (contrarian)* idea (Bondt and Thaler 1985), which means that good (poor)-performing assets will perform poor (good) in the subsequent periods. Thus, follow the loser's approaches often are characterized by transferring capital from poor-performing assets (losers) to good-performing assets (winners). Although this principle is heavily investigated in finance journals, it has not been widely disseminated in the topic of online portfolio selection. However, some algorithms do follow this principle. One famous example is the CRP benchmark. Moreover, Cover's UP, which buys and holds CRP strategies, can also be viewed as follows the loser approach from the underlying stocks' perspective, while we categorize it as follow the winner from the experts' perspective.

This chapter is organized as follows. Section 5.1 illustrates the mean reversion idea, which is the key underlying the "follow the loser" principle. Section 5.2 introduces a representative strategy in this category, or the Anticor strategy. Finally, Section 5.3 summarizes the follow the loser principle.

5.1 Mean Reversion

Besides the momentum-related idea that assumes that the stock price will continue its previous trend, there exists another different idea, or the mean reversion (contrarian) idea, which assumes that the assets' prices will revert to their means. Thus, the follow the loser algorithms will transfer the wealth from outperforming assets to underperforming assets.

This section illustrates a simple but convincing example to show the mean reversion idea. Consider a fluctuating market with two assets (A,B), and the price relative sequence is $\left(\frac{1}{2}, 2\right)$, $\left(2, \frac{1}{2}\right)$, ..., where each asset is not going anywhere but actively moving within a range (Table 5.1). Obviously, in the long run, a market strategy cannot achieve any abnormal return since the cumulative wealth of each stock remains

Table 5.1 *Motivating example to show the mean reversion trading idea*

Period #	Market (A,B)	BCRP	BCRP Return	Adjusted Weights	Notes
1	$(1/2, 2)$	$(1/2, 1/2)$	$5/4$	$(1/5, 4/5)$	$B \longrightarrow A$
2	$(2, 1/2)$	$(1/2, 1/2)$	$5/4$	$(4/5, 1/5)$	$A \longrightarrow B$
3	$(1/2, 2)$	$(1/2, 1/2)$	$5/4$	$(1/5, 4/5)$	$B \longrightarrow A$
$\vdots$	$\vdots$	$\vdots$	$\vdots$	$\vdots$	$\vdots$

the same after $2n$ periods. However, BCRP in hindsight can achieve a growth rate of $\left(\frac{5}{4}\right)^n$ for a n-trading period.

Now let us analyze the BCRP's behaviors to show the underlying mean reversion trading idea (Table 5.1). Suppose the initial portfolio is $\left(\frac{1}{2}, \frac{1}{2}\right)$ and at the end of period 1, the close price adjusted portfolio distribution becomes $\left(\frac{1}{5}, \frac{4}{5}\right)$ and cumulative wealth increases by a factor of $\frac{5}{4}$. At the beginning of period 2, portfolio manager rebalances to initial portfolio $\left(\frac{1}{2}, \frac{1}{2}\right)$ by transferring the wealth from a better-performing asset (B) to a worse-performing asset (A). At the beginning of period 3, the wealth transfer with the mean reversion trading idea continues. Although the market strategy gains nothing, BCRP can achieve a growth rate of $\frac{5}{4}$ per period with the underlying mean reversion trading idea, which assumes that if one asset performs worse, it tends to perform better in the subsequent trading period. It actually gains profit via the volatility of the market, or so-called volatility pumping (Luenberger 1998, Chapter 15).

Though extensive studies in finance show that mean reversion is a plausible idea to be used in trading (Chan 1988; Poterba and Summers 1988; Lo and MacKinlay 1990; Conrad and Kaul 1998), its counterintuitive nature hides it from the OLPS community. While the "follow the winner" strategies are sound in theory, they often perform poorly when using real data, which will be shown in the empirical studies in Part IV. Perhaps the reason is that their momentum principle does not fit the real market, especially on the tested trading frequency (such as daily). It is thus natural to utilize the mean reversion idea in developing new strategies so as to boost the empirical performance.

5.2 Anticorrelation

Borodin et al. (2004) proposed a follow the loser strategy named an *Anticorrelation* (Anticor). Instead of making no distributional assumption like Cover's UP, Anticor assumes that the market follows the mean reversion principle. To exploit the property, it statistically makes bets on the consistency of positive lagged cross-correlation and negative autocorrelation.

Anticor adopts logarithmic price relatives (Hull 1997) in two specific market windows, that is, $\mathbf{y}_1 = \log(\mathbf{x}_{t-2w+1}^{t-w})$ and $\mathbf{y}_2 = \log(\mathbf{x}_{t-w+1}^{t})$. It then calculates a cross-correlation matrix between $\mathbf{y}_1$ and $\mathbf{y}_2$,

$$M_{cov}(i, j) = \frac{1}{w-1}(\mathbf{y}_{1,i} - \bar{\mathbf{y}}_1)^{\top}(\mathbf{y}_{2,j} - \bar{\mathbf{y}}_2)$$

$$M_{cor}(i, j) = \begin{cases} \frac{M_{cov}(i,j)}{\sigma_1(i) \times \sigma_2(j)} & \sigma_1(i), \sigma_2(j) \neq 0 \\ 0 & \text{otherwise} \end{cases}.$$

Following the mean reversion principle, Anticor transfers weights from the assets increased more to the assets increased less, and the corresponding amounts are adjusted by the cross-correlation matrix. In particular, if asset i increases more than asset j and they are positively correlated, Anticor claims a transfer from asset i to j with the amount equaling the cross-correlation ($M_{cor}(i, j)$) minus their negative auto-correlation ($\min\{0, M_{cor}(i, i)\}$ and $\min\{0, M_{cor}(j, j)\}$). Finally, these claims are normalized to keep the portfolio in the simplex domain.

With the mean reversion nature, it is difficult to obtain a useful regret bound for Anticor. Although heuristic and without theoretical guarantee, Anticor empirically outperforms all other strategies at the time. On the other hand, though Anticor obtains good performance, its heuristic nature cannot fully exploit mean reversion. Thus, exploiting the property via systematic learning algorithms is highly desired, which motivates one part of our research.

5.3 Summary

Although counterintuitive, the follow the loser principle is quite useful in obtaining a high cumulative return in the empirical studies. This may be attributed to the fact that many financial research studies have validated that the market behaviors follow the mean reversion principle. Thus, to better exploit the market, a trading strategy has to incorporate the market behaviors. We further propose three novel mean reversion-based algorithms in Chapters 9, 10, and 11, respectively.

Chapter 6

Pattern Matching

Besides follow the winner and follow the loser, another category utilizes both winners and losers, and it is based on pattern matching. This category mainly covers nonparametric sequential investment strategies, which guarantee an optimal growth of capital under minimal assumptions on the market, that is, stationary and ergodic of the financial time series. Based on nonparametric prediction (Györfi and Schäfer 2003), this category consists of several pattern matching–based investment strategies (Györfi et al. 2006, 2007, 2008; Li et al. 2011a). Note that in the data-mining communities, some researchers focus on detecting important signals or patterns in time series (Mcinish and Wood 1992; Berndt and Clifford 1994; Agrawal and Srikant 1995; Srikant and Agrawal 1996; Ting et al. 2006; Cañete et al. 2008; Du et al. 2009), which is beyond our discussion.

In general, the pattern matching–based approaches (Györfi et al. 2006) consist of two steps, that is, the sample selection and portfolio optimization steps. Suppose we are choosing a portfolio for period $t + 1$. First, the sample selection step selects a set C_t of similar historical indices, whose corresponding price relatives will be used to predict the next one. Then, each price relative vector $\mathbf{x}_i$, $i \in C_t$, is assigned a probability of P_i, $i \in C_t$. Existing methods often choose uniform probability $P_i = \frac{1}{|C_t|}$, where $|\cdot|$ denotes the cardinality of a set. Second, the portfolio optimization step learns an optimal portfolio based on the selected set, that is,

$$\mathbf{b}_{t+1} = \arg\max_{\mathbf{b} \in \Delta_m} U(\mathbf{b}, C_t),$$

where $U(\cdot)$ is a specified utility function, such as log utility. In case of an empty sample set, a uniform portfolio is adopted.

In this chapter, we concretize the sample selection step in Section 6.1 and the portfolio optimization step in Section 6.2. We finally combine the two steps to formulate specific online portfolio selection algorithms in Section 6.3. Based on the principle, we further proposed the correlation-driven nonparametric learning (CORN) algorithm in Chapter 8.

35

6.1 Sample Selection Techniques

The general idea of this step is to select similar samples from historical price relatives by comparing two preceding market windows. Suppose we are locating the price relative vectors that are similar to the next vector $\mathbf{x}_{t+1}$. The basic routine is to iterate all historic price relatives $\mathbf{x}_i$, $i = w + 1, \ldots, t$ and count $\mathbf{x}_i$ as one similar vector, if its preceding market window $\mathbf{x}_{i-w}^{i-1}$ is similar to the latest market window $\mathbf{x}_{t-w+1}^{t}$. A set C_t contains the indexes of similar price relatives. Note that the market window is a $w \times m$-matrix and the similarity is typically calculated on the concatenated $w \times m$-vectors. Algorithm 6.1 further illustrates the procedure.

Algorithm 6.1: Sample selection procedure $(C(\mathbf{x}_1^t, w))$.

Input: $\mathbf{x}_1^t$: Historical market sequence; w: window size.
Output: C_t: Index set of similar price relatives.

Initialize $C_t = \emptyset$;
if $t \leq w$ **then**
 | return;
end
for $i = w + 1, w + 2, \ldots, t$ **do**
 | **if** $\mathbf{x}_{i-w}^{i-1}$ *is similar to* $\mathbf{x}_{t-w+1}^{t}$ **then**
 | | $C_t = C_t \cup i$;
 | **end**
end

A nonparametric *histogram-based* sample selection (Györfi and Schäfer 2003) predefines a set of discretized partitions, partitions both the latest market window $(\mathbf{x}_{t-w+1}^{t})$ and historical market windows $(\mathbf{x}_{i-w}^{i-1}, i = w + 1, \ldots, t)$, and finally chooses price relatives $\mathbf{x}_i$ whose preceding market window $(\mathbf{x}_{i-w}^{i-1})$ is in the same partition as $\mathbf{x}_{t-w+1}^{t}$. In particular, given a partition $P = A_j$, $j = 1, 2, \ldots, d$, which discretizes $\mathbb{R}_+^m$ into d disjoint sets, and a corresponding discretization function $G(\mathbf{x}) = j$, we can define the similarity set as

$$C_H(\mathbf{x}_1^t, w) = \left\{ w < i < t + 1 : G(\mathbf{x}_{t-w+1}^{t}) = G(\mathbf{x}_{i-w}^{i-1}) \right\}.$$

Nonparametric *kernel-based* sample selection (Györfi et al. 2006) identifies the similarity set by evaluating the Euclidean distance between two market windows,

$$C_K(\mathbf{x}_1^t, w) = \left\{ w < i < t + 1 : \left\| \mathbf{x}_{t-w+1}^{t} - \mathbf{x}_{i-w}^{i-1} \right\| \leq \frac{c}{\ell} \right\},$$

where c and ℓ are thresholds used to control the number of similar samples.

Nonparametric *nearest neighbor-based* sample selection (Györfi et al. 2008) searches price relatives whose preceding market windows are within the k nearest neighbors of the latest market window, that is,

$$C_N(\mathbf{x}_1^t, w) = \left\{ w < i < t + 1 : \mathbf{x}_{i-w}^{i-1} \text{ is among the } k \text{ NNs of } \mathbf{x}_{t-w+1}^{t} \right\},$$

where k is a threshold parameter.

6.2 Portfolio Optimization Techniques

The second step of pattern matching–based approaches is to construct an optimal portfolio based on the sample set C_t. Two main principles are Kelly's (1956) capital growth portfolio and Markowitz's (1952) mean variance portfolio.

Györfi et al. (2006) proposed to figure out a *log-optimal* (Kelly) portfolio, based on similar price relatives, which clearly follows the capital growth portfolio theory. Given a sample set C_t, the log-optimal utility function is defined as

$$U_L(\mathbf{b}, C_t) = \mathbb{E}\{\log \mathbf{b} \cdot \mathbf{x} | \mathbf{x}_i, i \in C_t\} = \sum_{i \in C_t} P_i \log \mathbf{b} \cdot \mathbf{x}_i,$$

where P_i denotes the probability assigned to $\mathbf{x}_i$, $i \in C_t$. Györfi et al. (2006) assumed a uniform probability, thus equivalently,

$$U_L(\mathbf{b}, C_t) = \sum_{i \in C_t} \log \mathbf{b} \cdot \mathbf{x}_i. \tag{6.1}$$

Maximizing the above function results in a BCRP portfolio (Cover 1991) over the similar price relatives.

Györfi et al. (2007) introduced *semi-log-optimal* utility function, which approximates log utility in Equation 6.1 aiming to release its computational complexity; and Vajda (2006) presented corresponding theoretical analysis and proved its universality. The semi-log-optimal utility function is defined as

$$U_S(\mathbf{b}, C_t) = \mathbb{E}\{f(\mathbf{b} \cdot \mathbf{x}) | \mathbf{x}_i, i \in C_t\} = \sum_{i \in C_t} P_i f(\mathbf{b} \cdot \mathbf{x}_i),$$

where $f(\cdot)$ is the second-order Taylor expansion of $\log z$ with respect to $z = 1$, that is,

$$f(z) = z - 1 - \frac{1}{2}(z - 1)^2.$$

Györfi et al. (2007) adopted a uniform probability of P_i, thus, equivalently,

$$U_S(\mathbf{b}, C_t) = \sum_{i \in C_t} f(\mathbf{b} \cdot \mathbf{x}_i).$$

Ottucsák and Vajda (2007) proposed a *Markowitz-type* utility function, which further generalizes the semi-log-optimal strategy. The basic idea is to trade off between portfolio mean and variance, which is similar to Markowitz's mean variance theory. To be specific, its utility function is defined as

$$U_M(\mathbf{b}, C_t) = \mathbb{E}\{\mathbf{b} \cdot \mathbf{x} | \mathbf{x}_i, i \in C_t\} - \lambda \text{Var}\{\mathbf{b} \cdot \mathbf{x} | \mathbf{x}_i, i \in C_t\}$$
$$= \mathbb{E}\{\mathbf{b} \cdot \mathbf{x} | \mathbf{x}_i, i \in C_t\} - \lambda \mathbb{E}\{(\mathbf{b} \cdot \mathbf{x})^2 | \mathbf{x}_i, i \in C_t\} + \lambda(\mathbb{E}\{\mathbf{b} \cdot \mathbf{x} | \mathbf{x}_i, i \in C_t\})^2,$$

where λ is a trade-off parameter. In particular, simple numerical transformations show that the semi-log-optimal portfolio is one special case of this utility function.

To solve the problem with transaction costs, Györfi and Vajda (2008) proposed a *GV-type* utility function* by incorporating the transaction costs (Gyorfi and Walk 2012),

$$U_T(\mathbf{b}, C_t) = \mathbb{E}\{\log \mathbf{b} \cdot \mathbf{x} + \log w(\mathbf{b}_t, \mathbf{b}, \mathbf{x}_t)\},$$

where $w(\cdot) \in (0, 1)$ is the transaction cost factor, which represents the remaining proportion after transaction costs. With a uniform probability assumption, it is equivalent to calculate:

$$U_T(\mathbf{b}, C_t) = \sum_{i \in C_t} (\log \mathbf{b} \cdot \mathbf{x}_i + \log w(\mathbf{b}_t, \mathbf{b}, \mathbf{x}_t)).$$

In any of the above procedures, if the similarity set is non-empty, we can obtain an optimal portfolio based on the similar price relatives and their assumed probability. In the case of an empty set, we can choose either a uniform portfolio or the last portfolio.

6.3 Combinations

Finally, let us combine the two steps and describe specific algorithms in the pattern matching–based approach. Table 6.1 summarizes all existing combinations.

One default utility function is the log-optimal function. Györfi and Schäfer (2003) introduced the *nonparametric histogram-based log-optimal* investment strategy (B^H), which combines the histogram-based sample selection and log-optimal utility function. Györfi et al. (2006) presented the *nonparametric kernel-based log-optimal* investment strategy (B^K), which combines the kernel-based sample selection and log-optimal utility function. Györfi et al. (2008) proposed the *nonparametric nearest neighbor log-optimal* investment strategy (B^{NN}), which combines the nearest neighbor sample selection and log-optimal utility function.

Besides the log-optimal utility function, several algorithms using different utility functions have been proposed. Györfi et al. (2007) proposed the *nonparametric kernel-based semi-log-optimal* investment strategy (B^S) by combining the kernel-based sample selection and semi-log-optimal utility function, which greatly eases

Table 6.1 *Pattern matching–based approaches: sample selection and portfolio optimization*

Portfolio Optimization	Sample Selection Techniques		
	Histogram	Kernel	Nearest Neighbor
Log-optimal	B^H: $C_H + U_L$	B^K: $C_K + U_L$	B^{NN}: $C_N + U_L$
Correlation-driven	—	CORN	—
Semi-log-optimal	—	B^S: $C_K + U_S$	—
Markowitz-type	—	B^M: $C_K + U_M$	—
GV-type	—	B^{GV}: $C_K + U_R$	—

Note: —, no algorithm in the combinations.

---

*Algorithm 2 in Györfi and Vajda (2008).

the computation of B^K. Ottucsák and Vajda (2007) proposed the *nonparametric kernel-based Markowitz-type* investment strategy (B^M) by combining the kernel-based sample selection and Markowitz-type utility function. Györfi and Vajda (2008) proposed the *nonparametric kernel-based GV-type* investment strategy (B^{GV}) by combining the kernel-based sample selection and GV-type utility function to select portfolios in case of nonzero transaction costs.

6.4 Summary

This chapter summarizes the pattern matching–based principle, which mainly includes pattern-matching and portfolio optimization steps. Empirically, these algorithms exploit recurring patterns over the history and produce good empirical performance. One of its key problems is to identify the recurring patterns, which leads to our CORN strategy in Chapter 8.

Chapter 7

Meta-Learning

Another research topic in online portfolio selection (OLPS) is *meta-learning*, or *meta-algorithms* (MAs) (Das and Banerjee 2011), which is closely related to expert learning (Cesa-Bianchi and Lugosi 2006). This is directly applicable to the "fund of fund" (FOF),* which delegates portfolio capital to other funds. In general, MA defines several base experts, each of which is equipped with strategies from the same strategy class or different classes, or even MAs. Each expert outputs a portfolio vector, and MA combines these portfolios to form a final portfolio, which is used for rebalance. The whole system can achieve the best performance among the experts in hindsight, which thus is desired for some nonuniversal algorithms. MAs are similar to Cover's UP algorithm in the follow the winner approach; however, they are proposed to handle different classes of experts, among which UP's CRP becomes a special case. On the one hand, MAs can be used to smooth the final performance with respect to all experts, especially when base experts are sensitive to certain environments/parameters. On the other hand, combining universal algorithms and heuristic algorithms, which is not easy to obtain a theoretical regret bound, can provide the universality property for the whole system. Finally, MAs can be applied to all existing approaches and thus have much broader areas of application.

7.1 Aggregating Algorithms

Though BCRP is optimal for an independent and identically distributed (i.i.d.) market, which is often suspected in real markets, the optimal portfolio may not belong to CRP. Several algorithms have been proposed to track a different set of experts. The base experts in this category belong to a special class rather than complex experts from multiple classes.

Vovk and Watkins (1998) applied the *aggregating algorithm* (AA) (Vovk 1990, 1997, 1999, 2001) to the OLPS task, of which Cover's UP is a special case. The general setting for AA is to define a countable or finite set of base experts and sequentially allocate the resource among multiple base experts to achieve a good performance that

---

*FOF selects portfolios on different fund managers, rather than on assets. For example, an FOF manager may evenly split his fund, and put one part to fund A and the other to Fund B.

is no worse than any fixed combination of underlying experts. Its portfolio update formula (Vovk and Watkins 1998, Algorithm 1) for OLPS is

$$\mathbf{b}_{t+1} = \frac{\int_{\Delta_m} \mathbf{b} \prod_{i=1}^{t-1} (\mathbf{b} \cdot \mathbf{x}_t)^{\eta} P_0(d\mathbf{b})}{\int_{\Delta_m} \prod_{i=1}^{t-1} (\mathbf{b} \cdot \mathbf{x}_t)^{\eta} P_0(d\mathbf{b})}.$$

As a special case, Cover's UP corresponds to AA with uniform prior distribution and $\eta = 1$.

Several further algorithms have been proposed. Singer (1997) proposed the *switching portfolios* (SP), which switches among a set of strategies handling different regimes. The author proposed two switching schemes, both of which assume the duration of base strategies is geometrically distributed. While the first strategy assumes a fixed distribution, the second assumes that the distribution is dynamically changing. The authors further presented the lower bound of its logarithmic wealth with respect to the best switching regime. Empirical evaluations show that SP can outperform UP, EG, and BCRP.

Levina and Shafer (2008) proposed the *Gaussian random walk* strategy, which switches among base experts according to a Gaussian distribution. Kozat and Singer (2007) extended SP to piecewise fixed fraction strategies, which partitions the periods into different segments and transits among these segments. Kozat and Singer (2008) extended Kozat and Singer (2007) to the cases of transaction costs. Kozat and Singer (2009, 2010) further generalized to sequential decision problems. Kozat et al. (2008) proposed another piecewise universal portfolio selection strategy via context trees, and Kozat et al. (2011) also generalized to sequential decision problems via tree weighting.

SP adopts the notion of regime switching (Hamilton 1994, 2008), which seems to be more plausible than an i.i.d. market assumption. Regime switching is also applied to some state-of-the-art trading strategies (Hardy 2001). However, existing geometrical and Gaussian distributions do not seem to fit the market well, which leads to other possible distributions that can fit the markets better.

7.2   Fast Universalization

Akcoglu et al. (2005) proposed *fast universalization* (FU), which extends Cover's (1991) UP from a parameterized CRP class to a wide class of investment strategies, including trading strategies operating on a single stock and portfolio strategies allocating wealth among the whole market. FU's basic idea is to evenly split the wealth among base experts, let these experts operate on their own, and finally pool their wealth. FU's update is the same as that of Cover's UP, and it also asymptotically achieves a growth rate that equals that of an optimal fixed convex combination of base experts. In cases in which all experts are CRPs, FU would downgrade to Cover's UP.

Formally, FU's investment can be described as

$$\mathbf{b}_t = \frac{\int_{\mathcal{W}} S_t(w) R_t(w) d\mu(w)}{\int_{\mathcal{W}} R_t(w) d\mu(w)}, \tag{7.1}$$

where $R_0(w) = 1$ for the $w \in \mathcal{W}$. Note the mean is the same as Cover's UP, which equally splits the money among different strategies and lets them run.

Besides the universalization in the continuous parameter space, various discrete buy-and-hold combinations have been adopted by various existing algorithms. Rewriting Cover's UP in its discrete form, the update can be straightforwardly obtained. For example, Borodin et al. (2004) adopted the BAH strategy to combine Anticor experts with respect to a finite number of window sizes (or parameters). Moreover, all pattern matching–based approaches adopted BAH to combine their underlying experts, also with a finite number of window sizes (or parameters).

7.3 Online Gradient and Newton Updates

Das and Banerjee (2011) proposed two meta-optimization algorithms, named *online gradient update* (OGU) and *online Newton update* (ONU), which are extended from *exponential gradient* (EG) and *online Newton step* (ONS), respectively. Since their updates and proofs are similar to their precedents, we ignore their updates. Theoretically, OGU and ONU can achieve the same growth rate as the optimal convex combination of underlying experts. Particularly, if any base expert is universal, then the final system enjoys the universality property. This property is useful, as an MA can combine a heuristic algorithm and a universal algorithm, and the final system can enjoy both superior heuristic performance and the universality property.

7.4 Follow the Leading History

Hazan and Seshadhri (2009) proposed the *follow the leading history* (FLH) algorithm for changing environments. FLH can incorporate various universal base experts, such as the ONS algorithm. Its basic idea is to maintain a working set of finite experts, which are dynamically added in and dropped out, and allocate the weights among some active working experts with an MA, for example, the Herbster–Warmuth algorithm (Herbster and Warmuth 1998). Different from other MAs with all experts operating from the beginning, FLH adopts experts starting from different periods. Theoretically, FLH based on universal algorithms is also universal, and empirically, FLH equipped with ONS can significantly outperform ONS.

7.5 Summary

Meta-learning is another widely discussed principle in the research of online portfolio selection (OLPS). It derives from base algorithms but treats these experts as the underlying assets. Thus, from this aspect, meta-algorithms (MAs) can be widely applied to all strategies discussed in previous chapters. We are interested in this principle because practical trading systems usually contain multiple strategies, and meta-learning can be used to combine these strategies in an effective way.

Part III

Algorithms

Chapter 8

Correlation-Driven Nonparametric Learning

As described in Part II, several approaches have been proposed to select portfolios from financial markets. The pattern matching–based approach, which is intuitive in nature, can achieve best performance at the present time. However, one key challenge to this approach is to effectively locate a set of trading days whose price relative vectors are similar to the coming one. As detailed in Section 6.1, existing strategies often adopt Euclidean distance to measure the similarity between two preceding market windows. Euclidean distance can somehow measure the similarity; however, it simply considers the neighborhood of the latest market windows and ignores the linear or nonlinear relationship between two market windows, which is important for price relative estimation. In this chapter, we propose to exploit similar patterns via a correlation coefficient, which effectively measures the linear relationship, and further propose a novel pattern matching–based online portfolio selection algorithm "CORrelation-driven Nonparametric learning" (CORN) (Li et al. 2011a). The proposed CORN algorithm can better locate a similarity set, and thus can output portfolios that are more effective than existing pattern matching–based strategies. Moreover, we also proved CORN's universal consistency,* which is a nice property for the pattern matching–based algorithms. Further, in Part IV, we will extensively evaluate the algorithm on several real stock markets, where the encouraging results show that the proposed algorithm can easily beat both market index and best stock substantially (without or with small transaction costs) and also surpass a variety of the state-of-the-art techniques significantly.

This chapter is organized as follows. Section 8.1 motivates the proposed correlation metric for selecting similarity sets. Section 8.2 details the ideas of the proposed online portfolio selection algorithm, and then Section 8.3 illustrates the proposed algorithms. Section 8.4 proves CORN's universal consistency and further analyzes the proposed algorithms. Finally, Section 8.5 summarizes this chapter and indicates future directions.

---

*This property is missing in Li et al. (2011a).

8.1 Preliminaries

8.1.1 Motivation

One main idea of existing approaches is to optimize portfolios by mining similar patterns and information from historical market sequences. Anticor (Borodin et al. 2004) attempts to find statistical relations between pairs of stocks, such as positive auto covariance and negative cross-covariance, while pattern matching–based strategies (Györfi et al. 2006, 2008) try to discover similar appearances among historical markets. Though successful in mining statistical relations among stocks, Anticor ignores market movements, which are crucial for a portfolio selection task. Moreover, Anticor is heuristic in nature, which could lead to suboptimal solutions. On the other hand, existing pattern matching–based strategies (Györfi et al. 2006, 2008) rely on Euclidean distance to measure the similarity between two market windows. Though their empirical performance is excellent, the Euclidean distance cannot exploit the directional information between the two market windows. Therefore, it may detect some useful price relatives, but often includes some potentially useless or even harmful price relatives and excludes many beneficial price relatives. Such a similarity set will finally weaken the following portfolio optimization step, resulting in less effective portfolios.

To better understand the drawbacks of Euclidean distance in measuring the similarity between two market windows, we give a motivating example in Figure 8.1. Let us assume that all market windows consist of two price relatives, such as a market of one asset and the window size is two, or a market with two assets and the window size equals one. Let the latest market window for the t-th period be $\mathbf{x}_{t-2}^{t-1} = (1.10, 1.20)$. Clearly, $\mathbf{x}_{t-2}^{t-1}$ shows an increasing trend, and we aim to locate similar market windows that also show increasing trends. Suppose we have three possible pairs of market windows: A1: (0.90, 0.80), A2: (0.80, 0.90); B1: (1.2, 1.1), B2: (1.1, 1.2); C1: (1.4, 1.3), C2: (1.3, 1.4). Note that in a long-only portfolio, relative trends, rather than absolute trends, determine the allocations of capital.* For example, although A2 contains two decreasing price relatives (both 0.90 and 0.80 are less than 1), the market sequence is relatively increasing (0.90 > 0.80). In case that the vectors contain two assets, for the recent market window x_{t-2}^{t-1}, it is better to allocate more capital on the second asset (1.20 > 1.10), which is also the case in A2. However, this is not the case in B1 or C1, though their absolute price relatives are all increasing.[†] Among the three pairs, A2, B2, and C2 show increasing trends, while A1, B1, and C1 show decreasing trends. Thus, a good similarity measure should classify A2, B2, and C2 as similar appearances, which will benefit the next step, and A1, B1, and C1 as dissimilar appearances, which will harm the subsequent portfolio optimization step.

Now let us classify these market sequences via a Euclidean distance measure with a radius of 0.2,[‡] that is, $\left\| \mathbf{x}_{i-2}^{i-1} - \mathbf{x}_{t-2}^{t-1} \right\| \leq 0.2$. According to Figure 8.1c, a Euclidean

---

*In our problem setting, there are no cash or risk-free assets. In reality, a weaker constraint (e.g., at most, 90% of capital can be put in assets), may appear in mutual funds.

[†]Because their first asset is more favorable than the second one, which is different from the latest $\mathbf{x}_{t-2}^{t-1}$.

[‡]The radius is arbitrarily chosen to to limit the number of selected price relatives.

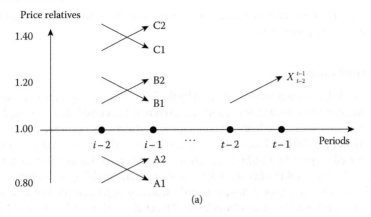

(a)

Sequence	Price relatives
$\mathbf{x}_{t-2}^{t-1}$	(1.1, 1.2)
A1	(0.9, 0.8)
A2	(0.8, 0.9)
B1	(1.2, 1.1)
B2	(1.1, 1.2)
C1	(1.4, 1.3)
C2	(1.3, 1.4)

(b)

$\mathbf{x}_{t-2}^{t-1}$	A1	A2	B1	B2	C1	C2
Euclidean distances	0.45	0.42	0.14	0	0.32	0.28
Similar? (Y/N)	N	N	Y	Y	N	N
Correlation coefficients	−1	1	−1	1	−1	1
Similar? (Y/N)	N	Y	N	Y	N	Y

(c)

Figure 8.1 *A motivating example to illustrate the limitation of Euclidean distance. (a) Market windows A1, A2, B1, B2, C1, C2, and* $\mathbf{x}_{t-2}^{t-1}$, *each of which contains two price relatives. (b) The price relative vectors. (c) The market windows via Euclidean distance and correlation coefficient.*

measure will classify B1 and B2 to the similarity set, since they are both located within the Euclidean ball of $\mathbf{x}_{t-2}^{t-1}$ (with a radius of 0.2). Such a classification is clearly suboptimal, as it includes harmful B1 and excludes beneficial A2 and C2. As a consequence of the imperfect similarity set, the subsequent portfolio optimization will considerably suffer from irrelevant or even harmful market windows (such as market window B1) and the neglect of beneficial market windows (such as market windows A2 and C2). This motivates us to overcome the limitation by exploring a more effective similarity measure.

8.2 Formulations

The proposed algorithm is mainly inspired by the idea of exploiting statistical correlations between two market windows, and also driven by the consideration of exploring the powerful nonparametric learning techniques to effectively optimize a portfolio.

Traditional portfolio selection methods in finance often try to estimate a target function based on past data and build portfolios based on the learned function. However, since the financial market is complex and accurate modeling of its movements is a difficult task, we adopt a nonparametric learning approach (or instance-based learning, or case-based learning) (Aha 1991; Aha et al. 1991; Cherkassky and Mulier 1998). Nonparametric learning makes no assumptions on data distribution (or market distribution), and it captures the knowledge from stored training data without building any target functions. In particular, at the beginning of every period, the proposed algorithm locates similar price relatives among all past price relatives, and then maximizes the expected multiplicative portfolio return directly based on the similar appearances. Without estimating any global functions of the market movements, the proposed algorithm estimates a target value of next price relative.

To overcome the limitation of Euclidean distance in mining historical market windows and the negligence of whole-market movements in all existing strategies, we propose to employ the *Pearson product–moment correlation coefficient*, which is an effective tool for measuring statistical linear relationships. Note that it measures the statistical correlations between market windows of all assets, rather than pairs of assets as Anticor does. Since market windows of all assets represent the whole-market movements in a period, they could be more effective to match the similar price relatives regarding the whole market.

Till now, we declare a *correlation-similar* set that contains historical trading days whose previous market windows are statistically correlated to the latest one, and formally define it as

$$C_t(w, \rho) = \left\{ w < i < t+1 \left| \frac{\text{cov}\left(\mathbf{x}_{i-w}^{i-1}, \mathbf{x}_{t-w+1}^t\right)}{\text{std}\left(\mathbf{x}_{i-w}^{i-1}\right)\text{std}\left(\mathbf{x}_{t-w+1}^t\right)} \geq \rho \right. \right\},$$

where w denotes the window size, $-1 \leq \rho \leq 1$ is a correlation coefficient threshold, $\text{cov}(A, B)$ denotes the covariance between market windows A and B, and $\text{std}(A)$ denotes the standard deviation of market window A. If either std term equals 0, that is, the market is of zero volatility in a specific window, we simply set the correlation coefficient to 0. In the above calculation, both matrix $\mathbf{x}_{i-w}^{i-1}$ and $\mathbf{x}_{t-w+1}^t$ are concatenated into $m \times w$-dimensional vectors, and we can obtain the univariate correlation coefficient between the two market windows.

The correlation coefficient measure distinguishes the proposed algorithm from previous nonparametric learning strategies, which measure the similarity via Euclidean distance. First, Euclidean distance only considers the magnitude between two market windows, while the proposed correlation coefficient measures their linear similarity, in both magnitude and direction. On the one hand, it considers the direction. For example, $\rho_1 = 0.8$ and $\rho_2 = -0.8$ intuitively correspond to equivalent magnitudes of linear dependence or similarity; however, they are in opposite directions, that is, the first market window is in the same trend with the latest market window, and the other is opposite. On the other hand, it also considers the magnitude. For example, $\rho_1 = 0.8$ and $\rho_2 = 0.2$ clearly indicate that the first market window is more suitable than the second one ($\rho_1 > \rho_2$). Thus, the correlation coefficient measure considers not only magnitude but also direction, which are appropriately balanced. With such linear dependence, we can better identify similar price relatives, thus leading to superior performance. Euclidean distance may also be used to measure directional information indirectly, for example, by using the slope of two centralized points. However, such a method only measures the directional information but ignores their magnitude.

Second, to calculate the univariate correlation coefficient, we will calculate the arithmetic mean of both $m \times w$-dimensional vectors. This mean return is uniformly distributed among m assets over w periods, which is in essence the market strategy. As a result, the mean return actually reflects the whole-market movements during the window. The correlation coefficient measures the linear dependency between two market windows, whose means represent the whole-market movements. This, therefore, distinguishes the proposed strategy from Anticor strategy and existing nonparametric learning strategies, which ignore the whole market.

Now let us return to the preceding motivating example and select a similarity set via correlation coefficient metric, with a threshold of 0. Figure 8.1(c) clearly shows that the metric can correctly classify these market windows, whose results are identical to our intuitive analysis. In particular, A2, B2, and C2 are classified as similar, and A1, B1, and C1 are classified as dissimilar. Note that our example is extremely straightforward and thus results in extreme values (either $+1$ or -1), which is not always the case.

8.3 Algorithms

Next, we present the proposed CORN algorithm, which exploits the *correlation-similar* set in optimizing portfolios for actively rebalancing.

We start by defining a set of $W \times P$ experts, each expert indexed by (w, ρ), that is,

$$\{\mathcal{E}(w, \rho) : w \geq 1, -1 \leq \rho \leq 1\},$$

and W represents the maximum window size and P represents the number of correlation coefficient thresholds. Each expert $\mathcal{E}(w, \rho)$ represents a CORN expert learning algorithm and outputs a portfolio, denoted as $\mathcal{E}(w, \rho) = \mathbf{b}(w, \rho)$.

As summarized in Algorithm 8.1, a CORN expert learning algorithm consists of two major steps. The first step, as illustrated in Section 8.2, is to locate a correlation-similar set via the correlation coefficient metric, and the second step is to obtain an optimized portfolio that can maximize the expected return, which is the main target of our research. After calculating the *correlation-similar* set $C_t(w, \rho)$ at the end of

---

Algorithm 8.1: CORrelation-driven Nonparametric expert: CORN (w, ρ).

Input: w: Window size; ρ: Correlation coefficient threshold; $\mathbf{x}_1^t$: Historical market sequence; t: Index of current period.

Output: $\mathbf{b}_{t+1}$: Expert's portfolio for period $t + 1$.

begin

    Initialize $C_t = \emptyset$, $\mathbf{b}_{t+1} = \left(\frac{1}{m}, \ldots, \frac{1}{m}\right)$;

    if $t \le w$ **then**

        return $\mathbf{b}_{t+1}$;

    end

    for $i = w+1, w+2, \ldots, t$ **do**

        if corrcoef $(\mathbf{x}_{i-w}^{i-1}, \mathbf{x}_{t-w+1}^{t}) \ge \rho$ **then**

            $C_t = C_t \cup i$;

        end

    end

    if $C_t \ne \emptyset$ **then**

        Search for an optimal portfolio: $\mathbf{b}_{t+1} = \arg \max_{\mathbf{b} \in \Delta_m} \prod_{i \in C_t} (\mathbf{b} \cdot \mathbf{x}_i)$;

    end

    return $\mathbf{b}_{t+1}$;

end

---

period t, we propose to learn an optimal portfolio following the idea of BCRP (Cover 1991), which maximizes the expected multiplicative return over the sequence of similar price relatives, that is,

$$\mathbf{b}_{t+1}(w, \rho) = \arg \max_{\mathbf{b} \in \Delta_m} \prod_{i \in C_t(w, \rho)} (\mathbf{b} \cdot \mathbf{x}_i), \tag{8.1}$$

where Δ_m represents an m-dimensional simplex. In case that $C_t(w, \rho)$ is empty (especially for a large ρ value), we will simply adopt uniform portfolio $\left(\frac{1}{m}, \ldots, \frac{1}{m}\right)$. Note that the *correlation-similar* set usually contains a large number of correlated price relatives. If one similar price relative vector has occurred frequently in history, it will also appear multiple times in the *correlation-similar* set. In other words, Equation 8.1 has more or less considered the occurrence/confidence of the correlated price relative vectors, which would avoid simply taking an extreme case in history.

We further combine all experts according to their historical performance $S_t(w, \rho)$ and a probability distribution function $q(w, \rho)$. Specifically, CORN combines experts' portfolios and calculates the final portfolio for period $t + 1$ as

$$\mathbf{b}_{t+1} = \frac{\sum_{w,\rho} q(w, \rho) S_t(w, \rho) \mathbf{b}_{t+1}(w, \rho)}{\sum_{w,\rho} q(w, \rho) S_t(w, \rho)}, \tag{8.2}$$

where $\mathbf{b}_{t+1}(w, \rho)$ represents the portfolio computed by expert $\mathcal{E}(w, \rho)$ and $S_t(w, \rho)$ represents its historical performance. For an individual expert, the higher its historical return, the higher its weight assigned in the final portfolio.

After releasing the price relative vector of $\mathbf{x}_{t+1}$, CORN updates the cumulative wealth,

$$S_{t+1} = S_t \times (\mathbf{b}_{t+1} \cdot \mathbf{x}_{t+1}).$$

For the underlying experts, CORN updates their cumulative wealth,

$$S_{t+1}(w, \rho) = S_t(w, \rho) \times (\mathbf{b}_{t+1}(w, \rho) \cdot \mathbf{x}_{t+1}),$$

where $S_t(w, \rho)$ represents the cumulative wealth achieved by expert $\mathcal{E}(w, \rho)$ till period t.

Therefore, it is straightforward that the cumulative wealth achieved by the proposed CORN strategy after n periods is equivalent to a q-weighted sum of all experts' returns,

$$S_n = \sum_{w,\rho} q(w, \rho) S_n(w, \rho). \tag{8.3}$$

Clearly, the final cumulative return is affected by all underlying experts, and the portions of contributions made by each expert are determined by the predefined distribution $q(w, \rho)$ and expert's performance $S_n(w, \rho)$.

Ideally, indexed by (w, ρ), we can choose CORN experts such that they cover all possible parameter settings, thus eliminating their effects. However, the computational cost of such a combination is inhibitively high. To boost the efficiency, we can choose finite discrete dimensions of the parameters, that is, a specified number of (w, ρ) combinations.

The selection of experts also trades off an individual expert's performance and its computational time. First, Equation 8.3 clearly shows that each expert contributes to the final cumulative wealth by its performance; thus, choosing a worse expert may lower the final performance. Second, the mixture's computation time is generally the summation of all experts' individual time. In other words, choosing too many experts, which cost too much time, may affect its practical scalability.

In this study, we first adopted uniform combination, which chooses a uniform distribution of $q(w, \rho)$, and named it "CORN uniform combination" (CORN-U). Algorithm 8.2 shows the details of the proposed CORN-U algorithm. In particular, we assign the same weights to all CORN experts, although the weights can be adjusted if we have more information. Moreover, CORN-U only considers $P = 1$ and chooses a specific value of ρ.

The above uniform combination algorithm may include some poor experts, leading to the degradation of overall performance. To overcome such limitations, the second algorithm, "CORN top-K combination" (CORN-K), combines only the top K best experts. Algorithm 8.3 illustrates the proposed CORN-K algorithm. In particular, it chooses the top K experts with the highest historical returns and uniformly combines them. That is, the strategy assigns the set of top K experts a uniform distribution $q(w, \rho) = \frac{1}{K}$, while the weights assigned to other experts are simply set to 0. Moreover, for the proposed CORN-K algorithm, we define $P \geq 1$ associated experts, each of which has a different ρ value.

Algorithm 8.2: Online portfolio selection with CORN uniform algorithm (CORN-U).

Input: W: Maximum window size; ρ: Correlation coefficient threshold;
$\quad$ $\mathbf{x}_1^n = (\mathbf{x}_1, \ldots, \mathbf{x}_n)$: Historical market sequence.
Output: S_n: Final cumulative wealth.
begin
$\quad$ Initialize S_0 and W experts: $S_0 = 1$, $\mathbf{b}_1 = \frac{1}{m}\mathbf{1}$, $q(w, \rho) = \frac{1}{W}$,
$\quad$ $S_0(w, \rho) = 1$, $\mathbf{b}_1(w, \rho) = \frac{1}{m}\mathbf{1}$, $w = 1, \ldots, W$;
$\quad$ **for** $t = 1$ **to** n **do**
$\quad\quad$ Rebalance the portfolio to $\mathbf{b}_t$;
$\quad\quad$ Receive current price relatives: $\mathbf{x}_t$;
$\quad\quad$ Update the cumulative wealth: $S_t = S_{t-1} \times (\mathbf{b}_t \cdot \mathbf{x}_t)$;
$\quad\quad$ Update the experts' cumulative wealth:
$\quad\quad$ $S_t(w, \rho) = S_{t-1}(w, \rho) \times (\mathbf{b}_t(w, \rho) \cdot \mathbf{x}_t)$;
$\quad\quad$ Update next portfolio:
$\quad\quad$ **begin**
$\quad\quad\quad$ **for** $w = 1$ **to** W **do**
$\quad\quad\quad\quad$ CORN expert (Algorithm 8.1) finds a portfolio:

$$\mathbf{b}_{t+1}(w, \rho) = \text{CORN}(w, \rho)$$

$\quad\quad\quad$ **end**
$\quad\quad\quad$ Combine experts' portfolios:

$$\mathbf{b}_{t+1} = \frac{\sum_w q(w, \rho) S_t(w, \rho) \mathbf{b}_{t+1}(w, \rho)}{\sum_w q(w, \rho) S_t(w, \rho)}$$

$\quad\quad$ **end**
$\quad$ **end**
end

Remarks on Aggregation: Note that the aggregation or combination rule described in Equation 8.2 is a special case of the general concept of exponential weighting. For a learning parameter $\eta > 0$, put

$$\mathbf{b}_{t+1} = \frac{\sum_{w,\rho} q(w, \rho) e^{\eta \log S_t(w,\rho)} \mathbf{b}_{t+1}(w,\rho)}{\sum_{w,\rho} q(w, \rho) e^{\eta \log S_t(w,\rho)}}.$$

If $\eta = 1$, then one gets the rule in Equation 8.2. The proof of universal consistency of B^H, B^K, and B^{NN} works without any difficulties if $\eta \leq 1$. However, the exponential results are superior if η is much larger than 1, but there is no theoretical support for this phenomenon. The large η corresponds to CORN-K with $K = 1$ (i.e., this rule is the

---

Algorithm 8.3: Online portfolio selection with CORN top-K algorithm (CORN-K).

Input: W: Maximum window size; P: The number of correlation coefficient thresholds; K: The number for top K experts; $\mathbf{x}_1^n = (\mathbf{x}_1, \dots, \mathbf{x}_n)$: historical market sequence.

Output: S_n: Final cumulative wealth.

begin

Initialize S_0 and $W \times P$ experts: $S_0 = 1$, $\mathbb{P} = \left\{0, \frac{1}{P}, \dots, \frac{P-1}{P}\right\}$,

$q(w, \rho) = \frac{1}{W \times P}$, $S_0(w, \rho) = 1$, $w = 1, \dots, W$, $\rho \in \mathbb{P}$;

for $t = 1$ **to** n **do**

Rebalance the portfolio to $\mathbf{b}_t$;

Receive current price relatives: $\mathbf{x}_t$;

Update the cumulative wealth: $S_t = S_{t-1} \times (\mathbf{b}_t \cdot \mathbf{x}_t)$;

Update the experts' cumulative wealth:

$S_t(w, \rho) = S_{t-1}(w, \rho) \times (\mathbf{b}_t(w, \rho) \cdot \mathbf{x}_t)$;

Update experts' weights:

begin

Select top K experts $\mathbb{K} = \{(w, \rho)\}$ w.r.t. $S_t(w, \rho)$;

Set weights for the top K experts: $q(w, \rho) = \frac{1}{K}$, $(w, \rho) \in \mathbb{K}$;

Set zero weights for other experts: $q(w, \rho) = 0$, $(w, \rho) \notin \mathbb{K}$;

end

Update next portfolio:

begin

for $w = 1$ **to** W **do**

for $\rho \in \mathbb{P}$ **do**

CORN expert (Algorithm 8.1) finds a portfolio:

$$\mathbf{b}_{t+1}(w, \rho) = \text{CORN}(w, \rho)$$

end

end

Combine top K experts' portfolios:

$$\mathbf{b}_{t+1} = \frac{\sum_{w,\rho} q(w, \rho) S_t(w, \rho) \mathbf{b}_{t+1}(w, \rho)}{\sum_{w,\rho} q(w, \rho) S_t(w, \rho)}$$

end

end

end

---

follow the winner rule, which experts called follow the leader in the machine-learning literature) (Cesa-Bianchi and Lugosi 2006). It would be nice to prove or disprove that the follow the leader aggregation results in universally consistent strategies (i.e., asymptotically it is of growth optimal for any stationary and ergodic market process).

8.4 Analysis

In this section, we first analyze CORN's universal consistency with respect to the class of all ergodic process.

Theorem 8.1 The portfolio scheme CORN is universal with respect to the class of all ergodic processes such that $\mathbb{E}\{|\log X_j|\} < \infty$, for $j = 1, \ldots, m$.

Proof *The proof can be found in Appendix B.1.1.*

In the CORN expert learning procedure, there are two key parameters: the correlation coefficient threshold ρ and the window size w. Below, we analyze how they affect the algorithms.

As shown in the motivating example, the correlation coefficient threshold ρ is critical to a *correlation-similar* set. If ρ is negative, the *correlation-similar* set would contain some negatively correlated price relative vectors or irrelevant vectors. On the other hand, if ρ is too large, for example, $\rho \geq 0.5$, the *correlation-similar* set would neglect some positively correlated vectors. Since the *correlation-similar* set is crucial in selecting optimal portfolios, it would harm the learning performance if it either contains negatively correlated vectors/irrelevant vectors or discards positively correlated vectors. Empirically, we found that the optimal ρ value is often dataset dependent, but often close to 0, which will be verified in Section 13.3.1. Moreover, we note that CORN would degrade to a special case when $\rho \to 1$. As $\rho \to 1$, fewer market windows are highly positively correlated to the latest window. In the extreme case of $\rho = 1$, $C_t(w, \rho)$ becomes almost empty, which thus reduces to the uniform CRP strategy.*

Another key parameter for the CORN expert learning process is window size. Since the calculation of correlation coefficient treats market windows as a vector, the window size does not have a significant impact on the final portfolio. When certain experts give very bad performance, the final result tends to be relatively stable since the proposed combination methods (viz., CORN-U and CORN-K) will reduce the impact of these experts and thus provide a stable portfolio. We will numerically analyze the effect of window size in Section 13.3.1, which shows that the proposed combination can effectively smoothen the performance curve.

The simplicity and effectiveness of CORN raise a fundamental question: Is it reasonable to select a portfolio using only market price information?" While our goal is not to resolve the philosophical debates between fundamental and technical analysts, we believe this work goes a long way to provide empirical evidence endorsing the effectiveness of technical analysis. Moreover, note that the success of CORN depends on three basic assumptions that form the basis of most technical analysis methods, including: (i) market action discounts everything; (ii) price moves in trends; and (iii) history tends to repeat itself. The first point assumes that stock prices at any given time reflect everything that has or could affect a company, including fundamental factors. And the second and third points directly lead to our proposed

---

*This is not a general case, which depends on the initial portfolio and default values if a similarity set is empty.

CORN algorithm and existing pattern matching–based approach. All these assumptions allow us to construct a portfolio using only similar appearances of historical market prices, without considering other factors, either technical or fundamental.

8.5 Summary

This chapter proposed a novel "CORrelation-driven Nonparametric learning" (CORN) strategy for online portfolio selection, which effectively exploits the statistical correlations hidden in stock markets, and benefits from the exploration of powerful nonparametric learning techniques. The proposed CORN algorithm is simple in nature and easy to implement, and has parameters that are easy to set. It also enjoys the universal consistency property. Our empirical studies on real markets, in Part IV, show that CORN can substantially beat the market index and the best stock, and also consistently surpasses a variety of state-of-the-art algorithms.

Currently, the proposed CORN can capture the linear relationship between two market windows, and it is possible to further capture their nonlinear relationship. Although high return strategies are often associated with high risk, it would be more attractive to develop a strategy that can manage the risk properly without slashing too much return. As an extension to this work, we are currently developing such risk-limiting strategies for CORN. In future, we plan to investigate theoretical insights of the algorithm and examine its extensions to improve the performance with high transaction costs.

Chapter 9

Passive–Aggressive Mean Reversion

This chapter proposes a novel online portfolio selection (OLPS) strategy named "passive–aggressive mean reversion" (PAMR) (Li et al. 2012). Unlike traditional trend-following approaches, the proposed approach relies upon the mean reversion relation of financial markets. We are the first to devise a loss function that reflects the mean reversion principle. Further equipped with passive–aggressive online learning (Crammer et al. 2006), the proposed strategy can effectively exploit mean reversion. By analyzing PAMR's update scheme, we find that it nicely trades portfolio return with volatility risk and reflects the mean reversion principle. We conduct extensive numerical experiments in Part IV to evaluate the proposed algorithms on various real datasets. In most cases, the proposed PAMR strategy outperforms all benchmarks and almost all state-of-the-art strategies under various performance metrics. In addition to superior performance, the proposed PAMR runs extremely fast and thus is very suitable for real-life online trading applications.

This chapter is organized as follows. Section 9.1 briefly reviews the ideas of existing trend-following strategies and motivates the proposed strategy. Section 9.2 formulates the proposed PAMR strategy, and Section 9.3 derives the algorithms. Section 9.4 further analyzes and discusses the algorithms. Finally, Section 9.5 summarizes this chapter and indicates future directions.

9.1 Preliminaries

9.1.1 Related Work

One popular trading idea in reality is *trend following* or *momentum*, which assumes that historically outperforming stocks would still perform better than others in future. Some existing algorithms, such as EG and ONS, approximate the expected logarithmic daily return and logarithmic cumulative return, respectively, using historical price relatives. Though this idea is easy to understand and makes fortunes for many of the best traders and investors, trend following is hard to implement effectively. In addition, in the short term, the stock price relatives may not follow previous trends (Jegadeesh 1990; Lo and MacKinlay 1990).

Besides trend following, another widely adopted approach is *mean reversion* (Cover and Gluss 1986; Cover 1991; Borodin et al. 2004), which is also termed as *contrarian*. This approach stems from the CRP strategy (Cover and Gluss 1986), which rebalances to an initial portfolio every period. The idea behind this approach is that if one stock performs worse than others, it tends to perform better in the following periods. As a result, a contrarian strategy is characterized by the purchase of securities that have performed poorly and the sale of securities that have performed well, or, quite simply, "*Sell the winner, buy the loser*." According to Lo and MacKinlay (1990), the effectiveness of mean reversion is due to positive cross-autocovariances across securities. Among existing algorithms, CRP, UP,* and Anticor adopt this idea. However, CRP and UP passively revert to the mean, while empirical evidence from the Anticor algorithm (Borodin et al. 2004) shows that active reversion to the mean may better exploit the fluctuation of financial markets and is likely to obtain much higher profit. On the other hand, although Anticor actively reverts to the mean, it is a heuristic method based on statistical correlations. In other words, it may not effectively exploit the mean reversion property.

Pattern matching–based nonparametric learning algorithms (B^K, B^{NN}, and CORN, etc.) can identify many market conditions, including both mean reversion and trend following. However, when searching similar price relatives, they may locate both mean reversion and trend-following price relatives, whose patterns are essentially opposite, thus weakening the following maximization of expected cumulative wealth.

In summary, both trend following and mean reversion can generate profit in the financial markets, if appropriately used. In the following, we will propose an active mean reversion–based portfolio selection method. Though simple in update rules, it empirically outperforms the existing strategies in most back-tests† with real market data, indicating that it appropriately takes advantage of the mean reversion trading idea.

9.1.2 Motivation

The proposed approach is motivated by the CRP (Cover and Gluss 1986), which adopt the mean reversion trading idea. As shown in Chapter 5, the mean reversion principle has not been widely investigated for OLPS.

Another motivation of the proposed algorithm is that, in financial crisis, all stocks drop synchronously or certain stocks drop significantly. Under such situations, actively rebalance may be inappropriate since it puts too much wealth on "mine" stocks, such as Bear Stearns‡ during the subprime crisis. To avoid potential risk concerning such "mine" stocks, it is better to stick to a previous portfolio, which

---

*From the expert level, UP follows the winner. However, since its experts belong to CRP, it also follows the loser in stock level. In the preceding survey, we classify it following the expert level.

†Back-test refers to testing a trading strategy via historical market data.

‡Bear Stearns was a US company whose stock price collapsed in September 2008.

constitutes the CRP strategy. Here, the reason to choose a passive CRP strategy is that these "mine" stocks are usually known only in hindsight, thus identifying them a priori is almost impossible. Thus, to avoid suffering too much from such situations, the proposed approach alternates between "aggressive" and "passive" reversion depending on market conditions. The passive mean reversion avoids the high risk of aggressive mean reversion, which would put most wealth on these "mine" stocks.

In the following, we propose a novel trading strategy named "passive–aggressive mean reversion," or PAMR for short. On the one hand, the underlying assumption is that better-performing assets would perform worse than others in the next period. On the other hand, if the market drops too much, we would stop actively rebalancing portfolios to avoid certain "mine" stocks and their associated risk. To exploit these intuitions, we suggest adopting passive–aggressive (PA) online learning (Crammer et al. 2006), which was originally proposed for classification. The basic idea of PA is that it *passively* keeps the previous solution if the loss is zero, while it *aggressively* updates the solution whenever the suffering loss is nonzero.

We now describe the proposed PAMR strategy in detail. Firstly, if the portfolio period return is below a threshold, we will try to keep the previous portfolio such that it *passively* reverts to the mean to avoid potential "mine" stocks. Secondly, if the portfolio period return is above the threshold, we will actively rebalance the portfolio to ensure that the *expected* portfolio daily return is below the threshold, in the belief that the next price relatives will revert. This sounds a bit counterintuitive, but it is indeed reasonable, because if the price relative reverts, keeping the expected portfolio return below the threshold enables one to maintain a high portfolio return in the next period. Here, the expected portfolio return is calculated with respect to historical price relatives, for example, in our study, the last price relative (Helmbold et al. 1998).

To further illustrate that aggressive reversion to the mean can be more effective than a passive one, let us continue the example that has a market going nowhere but actively fluctuating. In such a market, the proposed strategy is much more powerful than best constant rebalanced portfolio (BCRP), a passive mean reversion trading strategy in hindsight, as shown in Table 9.1. As the motivating example shows, BCRP grows to $\left(\frac{5}{4}\right)^n$ for a n-trading period, while at the same time, PAMR grows to $\frac{5}{4} \times \left(\frac{3}{2}\right)^{n-1}$ (the details of the calculation/algorithm will be presented in the next section). We intuitively explain the success of PAMR below.

Assume the threshold for a PAMR update is set to 1, that is, if the portfolio period return is below 1, we do nothing but keep the existing portfolio. Our strategy begins with a portfolio $\left(\frac{1}{2}, \frac{1}{2}\right)$. For period 1, the return is $\frac{5}{4} > 1$. Then, at the beginning of period 2, we rebalance the portfolio such that an *approximate* portfolio return based on last price relatives is below the threshold of 1, and the resulting portfolio is $\left(\frac{2}{3}, \frac{1}{3}\right)$. As the mean reversion principle suggests, although we are building a portfolio performing below the threshold in the current period, we are actually maximizing the next portfolio return. As we can observe, the return for period 2 is $\frac{3}{2} > 1$. Then, following the same rule, we will rebalance the portfolio to $\left(\frac{1}{3}, \frac{2}{3}\right)$. As a result, in such a market, PAMR's growth rate is $\frac{5}{4} \times \left(\frac{3}{2}\right)^{n-1}$ for a n-period, which is superior to BCRP's $\left(\frac{5}{4}\right)^n$.

Table 9.1 *Motivating example to compare BCRP and PAMR*

		BCRP		PAMR		
Period #	Relatives	Portfolio	Return	Portfolio	Return	Notes
1	$(1/2, 2)$	$(1/2, 1/2)$	$5/4$	$(1/2, 1/2)$	$5/4$	Rebalance to $(2/3, 1/3)$
2	$(2, 1/2)$	$(1/2, 1/2)$	$5/4$	$(2/3, 1/3)$	$3/2$	Rebalance to $(1/3, 2/3)$
3	$(1/2, 2)$	$(1/2, 1/2)$	$5/4$	$(1/3, 2/3)$	$3/2$	Rebalance to $(2/3, 1/3)$
4	$(2, 1/2)$	$(1/2, 1/2)$	$5/4$	$(2/3, 1/3)$	$3/2$	Rebalance to $(1/3, 2/3)$
$\vdots$	$\vdots$	$\vdots$	$\vdots$	$\vdots$	$\vdots$	$\vdots$

Remarks on Motivations: Although the motivating example in Table 9.1 demonstrates the effectiveness of PAMR over BCRP, PAMR may not always outperform BCRP. In general, PAMR is an online algorithm, whereas BCRP is an optimal offline algorithm for i.i.d. markets (Cover and Thomas 1991, Theorem 15.3.1). Now, we discuss some possible situations where PAMR may fail to outperform BCRP.

Consider a special case where one stock crashes and the other explodes, for example, a market sequence of two stocks as $\left(\frac{1}{2}, 2\right), \left(\frac{1}{2}, 2\right), \ldots$. In this market, BCRP increases at an exponential rate of 2^n as it wholly invests in the second asset, while PAMR keeps a fixed wealth of $\frac{5}{4}$ over the trading period. Obviously, in such situation, PAMR performs much worse than BCRP, that is, PAMR's $\frac{5}{4}$ versus BCRP's 2^n over n periods. Though not shining in this example, PAMR still bounds its losses. Moreover, such a market, which violates the mean reversion assumption, is occasional, at least from the viewpoint of our empirical studies.

9.2 Formulations

Now we shall formally devise the proposed PAMR strategy for the OLPS task. PAMR is based on a loss function that exploits the mean reversion idea, which is our innovation, and is equipped with the PA online learning technique (Crammer et al. 2006).*

First of all, given a portfolio vector $\mathbf{b}$ and a price relative vector $\mathbf{x}_t$, we define an ϵ-insensitive loss function for the t-th period as

$$\ell_\epsilon(\mathbf{b}; \mathbf{x}_t) = \begin{cases} 0 & \mathbf{b} \cdot \mathbf{x}_t \leq \epsilon \\ \mathbf{b} \cdot \mathbf{x}_t - \epsilon & \text{otherwise} \end{cases}, \tag{9.1}$$

---

*In fact, with the loss function, we can adopt any learning methods to exploit the mean reversion property. We choose PA for its simplicity and effectiveness. Certainly, other learning techniques can be adopted, if the new method can provide some new insights.

where $\epsilon \geq 0$ is a sensitivity parameter that controls the mean reversion threshold. Since portfolio daily return fluctuates around 1,[*] we empirically choose $\epsilon \leq 1$ to buy underperforming assets. The ϵ-insensitive loss is zero when return is less than the threshold ϵ, and otherwise grows linearly with respect to portfolio return. For conciseness, let us use ℓ_ϵ^t to denote $\ell_\epsilon(\mathbf{b}; \mathbf{x}_t)$. By defining this loss function, we can distinguish the preceding two motivating cases.

Then, we will formulate the proposed strategy and will propose specific algorithms to solve them. Recalling that $\mathbf{b}_t$ denotes the portfolio vector for the period t, the first proposed method for PAMR is formulated as a constrained optimization.

Optimization Problem 1: PAMR

$$\mathbf{b}_{t+1} = \arg \min_{\mathbf{b} \in \Delta_m} \frac{1}{2} \|\mathbf{b} - \mathbf{b}_t\|^2 \quad \text{s. t.} \quad \ell_\epsilon(\mathbf{b}; \mathbf{x}_t) = 0. \tag{9.2}$$

The above formulation attempts to find an optimal portfolio by minimizing the deviation from last portfolio $\mathbf{b}_t$ if the constraint of zero loss is satisfied. On the one hand, the above approach *passively* keeps the last portfolio, that is, $\mathbf{b}_{t+1} = \mathbf{b}_t$, whenever the loss is zero, or the portfolio daily return is below the threshold ϵ. On the other hand, whenever the loss is nonzero, it *aggressively* updates the solution by forcing it to strictly satisfy the constraint, that is, $\ell_\epsilon(\mathbf{b}_{t+1}; \mathbf{x}_t) = 0$. Clearly, this formulation is able to address the two motivations.

Although the above formulation is reasonable to address our concerns, it may have some undesirable properties when noisy price relatives exist, which are common in real-world financial markets. For example, a noisy price relative in a trending sequence may suddenly change the portfolio in a wrong direction due to the aggressive update. To avoid such problems, we propose two variants of PAMR that are able to trade off between aggressiveness and passiveness. The idea of the two variants is similar to soft margin support vector machines by introducing some nonnegative slack variables into optimization. Specifically, for the first variant, we modify the objective function by introducing a term that scales linearly with respect to a slack variable ξ and formulate the following optimization.

Optimization Problem 2: PAMR-1

$$\mathbf{b}_{t+1} = \arg \min_{\mathbf{b} \in \Delta_m} \left\{ \frac{1}{2} \|\mathbf{b} - \mathbf{b}_t\|^2 + C\xi \right\} \quad \text{s.t.} \quad \ell_\epsilon(\mathbf{b}; \mathbf{x}_t) \leq \xi \text{ and } \xi \geq 0, \tag{9.3}$$

where C is a positive parameter to control the influence of the slack variable on the objective function. We refer to this parameter as an aggressiveness parameter similar to PA learning (Crammer et al. 2006) and call this variant "PAMR-1."

Instead of a linear slack variable, for the second variant, we modify the objective function by introducing a term that scales quadratically with respect to a slack variable ξ, which results in the following optimization problem.

---

[*]Here we use simple gross return, as defined in Section 9.2. Financial literature often adopts simple net return (Tsay 2002), which fluctuates around 0.

Optimization Problem 3: PAMR-2

$$\mathbf{b}_{t+1} = \arg\min_{\mathbf{b}\in\Delta_m} \left\{ \frac{1}{2}\|\mathbf{b}-\mathbf{b}_t\|^2 + C\xi^2 \right\} \quad \text{s.t.} \quad \ell_\epsilon(\mathbf{b};\mathbf{x}_t) \leq \xi. \tag{9.4}$$

We refer to this variant as "PAMR-2."

Remarks on Loss Function: In our loss function of Equation 9.1, we use the portfolio return $\mathbf{b}\cdot\mathbf{x}_t$, while it is possible to use log return $\log(\mathbf{b}\cdot\mathbf{x}_t)$ (Latané 1959).* With the log utility, optimization problems Equations 9.2 through 9.4 are all nonconvex and nonlinear, and thus difficult to solve. One way to solve them is to use log's first-order Taylor expansion at last portfolio and ignore higher order terms, that is,

$$\log(\mathbf{b}\cdot\mathbf{x}_t) \approx \log(\mathbf{b}_t\cdot\mathbf{x}_t) + \frac{\mathbf{x}_t}{\mathbf{b}_t\cdot\mathbf{x}_t}(\mathbf{b}-\mathbf{b}_t).$$

After the approximation, the nonlinear term becomes linear, and the optimization problems are thus convex and can be efficiently solved. However, such linear approximation may have some drawbacks. First of all, there is no way to justify the goodness of linear approximation. With log utility, the loss function is flat, then sharply rises and finally flattens out. While linear approximation is good in the two flat regimes, it is terrible at the point of nondifferentiability and subpar in the sharply rising region.

Moreover, linear approximation yields a upper bound on regret of log utility loss function. For the loss function in the form of Equation 9.1 without log utility or with log's linear approximation, the best possible regret, in a minimax sense, is at most $O(\sqrt{n})$ (Abernethy et al. 2009), while true log loss minimization algorithm can routinely achieve $O(\log n)$. However, our loss function is not a traditional loss function maximizing return (or minimizing the loss of $-\log\mathbf{b}\cdot\mathbf{x}_t$), but only a tool to realize mean reversion. Thus, the regret achieved using our loss function does not represent a regret about return, which may not be meaningful as traditional regret bound is. Anyway, though on empirical evaluations PAMR works well, anyone who cares about its theoretical aspects should be notified about the possible worse bound, which may not be elicited by the empirical evaluations.

Remarks on Formulations: Although our formulations mainly focus on portfolio return without explicitly dealing with risk (e.g., volatility of daily returns), the final algorithms can be nicely interpreted as certain trade-offs between risk and return, as discussed in Section 9.4. Such an interesting observation is further verified by our empirical evaluation, which shows that the proposed PAMR algorithms achieve good risk-adjusted returns in terms of two risk-related metrics (volatility risk and drawdown risk).

Similar to existing studies, our formulations avoid incorporating transaction cost, which simplifies and highlights PAMR's key ingredients. As shown in Sections 2.2 and 13.4, it is straightforward to evaluate the impact of transaction costs. In Chapter 13, we present results on both cases: with and without transaction costs. The results

---

*Empirically, log utility does not help much, since $\log(\mathbf{b}\cdot\mathbf{x}_t)$ and $\mathbf{b}\cdot\mathbf{x}_t$ are both small. However, theoretically, using log utility may help. We remark on it so as to attract theoretical interest from other researchers.

show that in most markets, the proposed algorithms work well without or even with moderate transaction costs.

9.3   Algorithms

We now derive the solutions for the three PAMR formulations using standard techniques from convex analysis (Boyd and Vandenberghe 2004) and present the proposed PAMR algorithms. Specifically, the following three propositions summarize their closed-form solutions.

Proposition 9.1 The solution to optimization problem 1 (PAMR) without considering the nonnegativity constraint ($\mathbf{b} \succeq 0$) is expressed as

$$\mathbf{b} = \mathbf{b}_t - \tau_t (\mathbf{x}_t - \bar{x}_t \mathbf{1}), \tag{9.5}$$

where $\bar{x}_t = \frac{\mathbf{x}_t \cdot \mathbf{1}}{m}$ denotes market return, and τ_t is computed as

$$\tau_t = \max \left\{ 0, \frac{\mathbf{b}_t \cdot \mathbf{x}_t - \epsilon}{\left\| \mathbf{x}_t - \bar{x}_t \mathbf{1} \right\|^2} \right\}. \tag{9.6}$$

Proof *The proof can be found in Appendix B.2.1.*

Proposition 9.2 The solution to optimization problem 2 (PAMR-1) without considering the nonnegativity constraint ($\mathbf{b} \succeq 0$) is expressed as

$$\mathbf{b} = \mathbf{b}_t - \tau_t \left(\mathbf{x}_t - \bar{x}_t \mathbf{1} \right),$$

where $\bar{x}_t = \frac{\mathbf{x}_t \cdot \mathbf{1}}{m}$ denotes market return, and τ_t is computed as

$$\tau_t = \max \left\{ 0, \min \left\{ C, \frac{\mathbf{b}_t \cdot \mathbf{x}_t - \epsilon}{\left\| \mathbf{x}_t - \bar{x}_t \mathbf{1} \right\|^2} \right\} \right\}. \tag{9.7}$$

Proof *The proof can be found in Appendix B.2.2.*

Proposition 9.3 The solution to optimization problem 3 (PAMR-2) without considering nonnegativity constraint ($\mathbf{b} \succeq 0$) is expressed as

$$\mathbf{b} = \mathbf{b}_t - \tau_t \left(\mathbf{x}_t - \bar{x}_t \mathbf{1} \right),$$

where $\bar{x}_t = \frac{\mathbf{x}_t \cdot \mathbf{1}}{m}$ denotes the market return, and τ_t is computed as

$$\tau_t = \max \left\{ 0, \frac{\mathbf{b}_t \cdot \mathbf{x}_t - \epsilon}{\left\| \mathbf{x}_t - \bar{x}_t \mathbf{1} \right\|^2 + \frac{1}{2C}} \right\}. \tag{9.8}$$

Proof *The proof can be found in Appendix B.2.3.*

Algorithm 9.1 details the proposed PAMR algorithms, and Algorithm 9.2 summarizes the OLPS procedure utilizing PAMR. Firstly, with no historical information, the initial portfolio is set to uniform $\mathbf{b}_1 = \left(\frac{1}{m}, \ldots, \frac{1}{m}\right)$. At the beginning of period t, we rebalance the portfolio following the decision made at the end of period $t - 1$. At the end of t-th period, the market reveals a stock price relative vector, which represents the market movements. Since both portfolio and price relatives are already known, the portfolio manager computes the portfolio daily return $\mathbf{b}_t \cdot \mathbf{x}_t$ and the loss $\ell_\epsilon(\mathbf{b}_t; \mathbf{x}_t)$ as defined in Equation 9.1. Then, we calculate an optimal step size τ_t based on last portfolio and stock price relatives. Given an optimal step size τ_t, we can update the portfolio for the next period. Finally, by projecting the updated portfolio into the simplex domain, we normalize the final portfolio.

Moreover, our algorithms have two key parameters, viz., the sensitivity parameter ϵ and the aggressiveness parameter C. In practice, their values could affect the performance of the proposed algorithms. To achieve a good performance in a specific market, the parameters have to be finely tuned. We will thoroughly examine the two parameters on real-life datasets and suggest their empirical selections in Section 13.3.2.

Algorithm 9.1: Passive–Aggressive Mean Reversion: PAMR(ϵ, C, $\mathbf{b}_t$, $\mathbf{x}_1^t$, t).

Input: $\epsilon \in [0, 1]$: sensitivity parameter; C: aggressiveness parameter; $\mathbf{b}_t$: current portfolio; $\mathbf{x}_1^t$: historical market sequence; t: index of current trading period.

Output: $\mathbf{b}_{t+1}$: Next portfolio.

begin

   Suffer loss: $\ell_\epsilon^t = \max\{0, \mathbf{b}_t \cdot \mathbf{x}_t - \epsilon\}$;

   Set parameters:

$$\tau_t = \begin{cases} \dfrac{\ell_\epsilon^t}{\left\|\mathbf{x}_t - \bar{x}_t \mathbf{1}\right\|^2} & \text{(PAMR)} \\[2em] \min\left\{C, \dfrac{\ell_\epsilon^t}{\left\|\mathbf{x}_t - \bar{x}_t \mathbf{1}\right\|^2}\right\} & \text{(PAMR-1)} \\[2em] \dfrac{\ell_\epsilon^t}{\left\|\mathbf{x}_t - \bar{x}_t \mathbf{1}\right\|^2 + \frac{1}{2C}} & \text{(PAMR-2)} \end{cases}$$

   Update portfolio:

$$\mathbf{b}_{t+1} = \mathbf{b}_t - \tau_t \left(\mathbf{x}_t - \bar{x}_t \mathbf{1}\right)$$

   Normalize portfolio:

$$\mathbf{b}_{t+1} = \arg\min_{\mathbf{b} \in \Delta_m} \|\mathbf{b} - \mathbf{b}_{t+1}\|^2$$

end

Algorithm 9.2: Online portfolio selection with PAMR.

Input: $\epsilon \in [0, 1]$: sensitivity parameter; C: aggressiveness parameter;
   $\mathbf{x}_1^n$: historical market sequence.
Output: S_n: final cumulative wealth.
begin

   Initialize $\mathbf{b}_1 = \left(\frac{1}{m}, \ldots, \frac{1}{m}\right)$, $S_0 = 1$;
   for $t = 1, \ldots, n$ **do**

      Rebalance the portfolio to $\mathbf{b}_t$;
      Receive stock price relatives: $\mathbf{x}_t = (x_{t,1}, \ldots, x_{t,m})$;
      Calculate the daily return and cumulative return: $S_t = S_{t-1} \times (\mathbf{b}_t^\top \mathbf{x}_t)$;
      Update the portfolio: $\mathbf{b}_{t+1} = \text{PAMR}(\epsilon, C, \mathbf{b}_t, \mathbf{x}_1^t, t)$;

   end

end

9.4 Analysis

To reflect the mean reversion trading idea, we are interested in analyzing PAMR's update rules, which mainly involve portfolio $\mathbf{b}_{t+1}$ and step size τ_t. In particular, we want to examine how the update rules are related to return and risk—the two most important concerns in a portfolio selection task.

First of all, we analyze the portfolio update rule for the three algorithms, that is,

$$\mathbf{b}_{t+1} = \mathbf{b}_t - \tau_t (\mathbf{x}_t - \bar{x}_t \mathbf{1}).$$

The step size τ_t is nonnegative, and $\bar{x}_t$ is mean return or market return. The $\mathbf{x}_t - \bar{x}_t \mathbf{1}$ represents stock abnormal returns with respect to the market on period t. We can further interpret it as a directional vector for the weight transfer. The negative sign before the term indicates that the update scheme is consistent with our motivation, that is, to transfer weights from outperforming stocks (with positive abnormal returns) to underperforming stocks (with negative abnormal returns).

It is interesting that the second part of the update,

$$\mathbf{a}_t = -\tau_t (\mathbf{x}_t - \bar{x}_t \mathbf{1}),$$

coincides with the general form (Lo and MacKinlay 1990, Eq. (1)) of return-based contrarian strategies (Conrad and Kaul 1998; Lo 2008), except a changing multiplier τ_t. This part represents an arbitrage (zero-cost) portfolio, since its elements always sum to 0, that is, $\mathbf{a}_t \cdot \mathbf{1} = 0$. Adding the arbitrage portfolio to the last portfolio, $\mathbf{b}_t$, results in the next portfolio. The long elements of the arbitrage portfolio ($a_{t,i} > 0$) increase the corresponding elements of the whole portfolio, and the short elements ($a_{t,i} < 0$) decrease the corresponding elements. Such an explanation is similar to the analysis in the last paragraph and connects PAMR's update with the general form of return-based contrarian strategies.

Besides, another important update is the step size τ_t calculated as Equations 9.6 through 9.8 for three PAMR methods, respectively. The step size τ_t adaptively controls the weights to be transferred by scaling the directional vector. One common term in τ_t is $\frac{\ell_\epsilon^t}{\|\mathbf{x}_t - \bar{x}_t \mathbf{1}\|^2}$. Its numerator denotes the ϵ-insensitive loss for period t, which equals the t-th portfolio return minus a mean reversion threshold, or zero. Assuming other variables are constant, if the return is high (low), it leads to a large (small) value of τ_t, which would aggressively transfer more (less) wealth from outperforming assets to underperforming assets. The denominator is essentially the market quadratic variability, that is, the number of assets times market variance of period t. In modern portfolio theory (Markowitz 1952), the variance of assets returns typically measures volatility risk for a portfolio. As indicated by the denominator, if the risk is high (low), the step size τ_t would be small (large). Consequently, the weight transfer made by the update scheme will be weakened (strengthened). This is consistent with our intuition that prediction would not be accurate in drastically dropping markets, and we opt to make less transfer to reduce risk. Moreover, PAMR-1 caps the step size by a constant C, while PAMR-2 decreases the step size by adding a constant $\frac{1}{2C}$ to its denominator. Both mechanisms can prevent drastic weight transfers in case of noisy price relatives, which is consistent with their motivations.

From the above analysis on the updates of portfolio and step size, we can conclude that PAMR nicely balances between return and risk and clearly reflects the mean reversion trading idea. To the best of our knowledge, such an important trade-off has only been considered by nonparametric kernel-based Markowitz-type strategy (Ottucsák and Vajda 2007). While the strategy trades off return and risk with respect to a set of similar historical price relatives, the proposed PAMR explicitly trades off return and risk with respect to last price relatives. This nice property distinguishes the proposed approach from most existing approaches that often cater to return, but ignore risk, and are therefore undesirable.

One objective for PAMR-1 and PAMR-2 is to prevent a portfolio from being affected too much from noisy price relatives, which might drastically change the portfolio. In this part, let us exemplify the benefits of PAMR's variants. Let $\mathbf{x}_t = (1.00, 0.01)$, whose second value is a noise, and $\mathbf{b}_t = (1, 0)$. Setting $\epsilon = 0.30$ and $C = 1.00$, we can calculate the next portfolio $\mathbf{b}_{t+1}$. This market sequence describes that certain stocks drop significantly, which is common during the financial crisis. Without tuning, PAMR would transfer a large proportion to the second asset. This can be verified by calculating PAMR's portfolio; in other words, PAMR calculates the update step size $\tau_t = 1.43$ and obtains the subsequent portfolio $\mathbf{b}_{t+1} = (0.29, 0.71)$. However, to avoiding such noises, a natural choice is to transfer less proportion to the second asset. On the other hand, PAMR-1 and PAMR-2 obtain the step sizes of $\tau_t = 1.00$ and $\tau_t = 0.71$, respectively, which are smaller than the original PAMR's. Accordingly, we obtain the next portfolios $\mathbf{b}_{t+1} = (0.50, 0.50)$ and $\mathbf{b}_{t+1} = (0.65, 0.35)$ for PAMR-1 and PAMR-2, respectively. Clearly, the variants transfer less wealth to the second asset than the original PAMR does. Thus, PAMR-1 and PAMR-2 suffer less from noisy price relatives, though they cannot completely avoid such suffering situations.

Finally, let us analyze PAMR's time complexity. Besides a normalization/ projection step (Step 7 in Algorithm 9.1), PAMR takes $O(m)$ per period. In our implementation, we adopt a linear projection method (Duchi et al. 2008),* which takes $O(m)$ per period. In total, the time complexity is $O(mn)$. Thus, PAMR has the same time complexity as the EG algorithm and is more superior to other methods. Linear time complexity enables the proposed algorithm to handle transactions in scenarios in which low latency is of crucial importance, such as high-frequency trading (Aldridge 2010).

9.5 Summary

In this chapter, we proposed a novel online portfolio selection (OLPS) strategy, passive–aggressive mean reversion (PAMR). Motivated by the idea of mean reversion and passive–aggressive online learning, PAMR either aggressively updates the portfolio following mean reversion, or passively keeps the previous portfolio. PAMR executes in linear time, making it suitable for online applications. We also find that its update scheme is based on the trade-off between return and volatility risk, which is ignored by most existing strategies. This interesting property connects the PAMR strategy with modern portfolio theory, which may provide further explanation from the aspect of finance.

The proposed algorithms are still far from perfect and may be improved in the following aspects. First of all, though the universality property may not be required in real investment, PAMR's universality is still an open question. Second, PAMR sometimes fails if mean reversion does not exist in the market components. Thus, it is crucial to locate asset sets exhibiting mean reversion. Finally, PAMR's formulations ignore transaction costs. Thus, directly incorporating the issue into formulations may improve PAMR's practical applicability.

---

*The precise MATLAB® routine *ProjectOntoSimplex* can be found on http://www.cs. berkeley.edu/~jduchi/projects/DuchiShSiCh08/

Chapter 10

Confidence-Weighted Mean Reversion

Empirical evidence (Borodin et al. 2004) shows that stock price relatives may follow the mean reversion property, which has not been fully exploited by existing strategies. Moreover, all existing online portfolio selection (OLPS) strategies only focus on the first-order information of a portfolio vector, though second-order information may also benefit a strategy. This chapter proposes a novel strategy named "confidence-weighted mean reversion" (CWMR) (Li et al. 2011b, 2013). Inspired by the mean reversion principle in finance and confidence-weighted (CW) online machine learning technique (Crammer et al. 2008; Dredze et al. 2008), CWMR models the portfolio vector as a Gaussian distribution, and sequentially updates the distribution following the mean reversion principle. Analysis of CWMR's closed form updates clearly reflects the mean reversion trading idea and the interaction of first-order and second-order information. Extensive experiments, in Part IV, on various real markets show that CWMR is able to effectively exploit the power of mean reversion and second-order information, and is superior to the state-of-the-art techniques.

This chapter is organized as follows. Section 10.1 motivates the proposed CWMR strategy. Section 10.2 formulates the strategy, and Section 10.3 derives the algorithms based on the formulations. Section 10.4 further analyzes the algorithms. Finally, Section 10.5 summarizes this chapter and indicates future directions.

10.1 Preliminaries

10.1.1 Motivation

The proposed method, similar to passive–aggressive mean reversion (PAMR), is based on the mean reversion trading idea, which, in the context of portfolio or multiple assets, implies that good-performing assets tend to perform worse than others in subsequent periods, and poor-performing assets are inclined to perform better. Thus, to maximize the next portfolio return, we could minimize the expected return with respect to today's price relatives since next price relatives tend to revert. This seems somewhat counterintuitive, but, according to Lo and MacKinlay (1990), the effectiveness of mean reversion is due to the positive cross-autocovariances across assets.

Besides the virtual example in Section 9.1.2, we empirically analyze real market data to show that mean reversion does exist.* Although measuring mean reversion in a single stock is well studied (Poterba and Summers 1988; Chaudhuri and Wu 2003; Hillebrand 2003), the study of mean reversion in a portfolio is rare. Since, in our formulation, the portfolio is long-only,[†] we focus on whether we can obtain a higher return than the market by investing on poor-performing assets.[‡] With a threshold δ, let A_t be the set of poor-performing stocks ($x_{t,i} < \delta$), B_t be the set of mean reversion (MR) stocks ($x_{t,i} < \delta$ & $x_{t+1,i} > 1$), C_t be the set of non–mean reversion (non–MR) stocks ($x_{t,i} < \delta$ & $x_{t+1,i} < 1$), and D_t be the set of remaining stocks ($x_{t,i} < \delta$ & $x_{t+1,i} = 1$). On period t, we calculate the percentage of a set U, which can be either A, B, C, or D, as $P_t(U) = |U_t|/|A_t|$, where $|\cdot|$ denotes the cardinality of a set, and the gain of uniform investment in the set as $G_t(U) = \sum_{i \in U_t} x_{t,i}/|U_t|$. For a total of n periods, we further calculate their average values as $\bar{P}(U) = \frac{1}{n-1}\sum_{t=1}^{n-1} P_t(U)$ and $\bar{G}(U) = \frac{1}{n-1}\sum_{t=1}^{n-1} G_t(U)$, respectively. In particular, we refer to the percentage of mean reversion stocks as $\bar{P}(B)$, and the gain of mean reversion stocks as $\bar{G}(B)$. To show whether buying poor-performing stocks is profitable, we calculate the average gain of uniform investment on poor-performing stocks, denoted as $\bar{G}(A)$, and the average gain of uniform investment in the whole market, denoted as $\bar{G}(Market)$. Table 10.1 gives the statistics on six real market daily datasets.[§] On the one hand, except for the DJIA dataset (please refer to Chapter 12 for details), mean reversion does exist ($\bar{P}(B) > \bar{P}(C)$),[¶] and uniform investment on poor-performing stocks provides a greater profit** than the market ($\bar{G}(A) > \bar{G}(Market)$). On the other hand, the test failed on the DJIA dataset, and in the following empirical evaluations, CWMR also failed badly on the dataset, which motivates our next proposed method in Chapter 11.

Moreover, all state-of-the-art approaches only exploit first-order information of a portfolio vector, while higher order information may also benefit the portfolio selection task (Harvey et al. 2010). Evidence (Chopra and Ziemba 1993) shows that in portfolio selection, errors in variance have about 5% impact on the objective value as errors in mean do. For simplicity, we exploit variance information while ignoring covariance information, which has a much smaller impact on the final objective value. To take advantage of both first- and second-order information, we adopt CW online learning (Crammer et al. 2008; Dredze et al. 2008), which was originally proposed for classification. CW's basic idea is to maintain a Gaussian distribution for a

---

*The test program and datasets will be available at http://stevenhoi.org/olps

[†]Long-only means if something is considered undervalued, managers would invest; if something is considered overvalued, managers would avoid it.

[‡]If short is allowed, we can also show whether shorting good-performing stocks provides a higher return.

[§]We list their details in Section 12.2. We empirically choose $\delta = 0.985$ on all datasets. As we have tested, other thresholds also release similar observations. For tests on other frequencies, please refer to Li et al. (2013).

[¶]This indicates a higher probability of reversion, but we have no theoretical guarantee for the criteria.

**The absolute return in the daily scale is relatively small. However, considering their net return, such a strategy makes much higher profit than the market does. Moreover, with compounding, such small absolute differences will result in huge differences over time.

Table 10.1 *Summary of mean reversion statistics on real markets*

Dataset	$\bar{P}(B)$	$\bar{G}(B)$	$\bar{P}(C)$	$\bar{G}(C)$	$\bar{P}(D)$	$\bar{G}(A)$	$\bar{G}(Market)$
TSE	42.89%	1.022370	41.63%	0.978395	15.48%	1.000598	1.000405
MSCI	54.19%	1.015737	45.05%	0.984046	0.76%	1.001107	1.000053
NYSE (O)	43.43%	1.021599	39.86%	0.981949	16.71%	1.002523	1.000620
NYSE (N)	47.87%	1.019624	43.19%	0.982050	8.93%	1.001644	1.000610
DJIA	48.54%	1.018545	50.57%	0.980843	0.90%	0.999398	0.999719
SP500	50.20%	1.020692	47.96%	0.980502	1.84%	1.000881	1.000488

classifier, and sequentially update the distribution similar to passive–aggressive (PA) learning (Crammer et al. 2006). Thus, CW learning can take advantage of both first- and second-order information of the classifier.

To address the above two concerns, we present a novel OLPS method named CWMR. To exploit the first- and second-order information of a portfolio vector, we model the portfolio vector as a Gaussian distribution, which is probably the most widely studied distribution and can satisfy our motivations. We do not consider higher orders and other distributions for their complexities. Then, we sequentially update the distribution following the mean reversion principle. On the one hand, we keep the previous distribution if the portfolio is profitable by using mean reversion. On the other hand, we move the distribution to a new distribution such that the new distribution is expected to make profit while keeping it close to the previous distribution. Different from CRP and Anticor, CWMR actively exploits the mean reversion property of financial markets with a powerful learning method. Moreover, compared with all existing algorithms, including PAMR, which only consider the first-order information, CWMR exploits both the first- and second-order information of a portfolio vector.

10.2 Formulations

We model $\mathbf{b}$ as a Gaussian distribution with mean $\mu \in \mathbb{R}^m$ and diagonal covariance matrix $\Sigma \in \mathbb{R}^{m \times m}$ with nonzero diagonal elements and zero for off-diagonal elements. The i-th element of μ represents the proportion of the i-th element. The i-th diagonal term of Σ stands for the confidence on the i-th proportion. The smaller the diagonal term, the higher the confidence we have in the corresponding μ.

At the beginning of period t, we figure out a $\mathbf{b}$ based on the distribution $\mathcal{N}(\mu, \Sigma)$, that is, $\mathbf{b} \sim \mathcal{N}(\mu, \Sigma)$. Then, after $\mathbf{x}_t$ is revealed, the wealth increases by a factor of $\mathbf{b}^\top \mathbf{x}_t$. It is straightforward that the return $D = \mathbf{b}^\top \mathbf{x}_t$ can be viewed as a random variable of the following univariate Gaussian distribution:

$$D \sim \mathcal{N}\left(\mu^\top \mathbf{x}_t, \mathbf{x}_t^\top \Sigma \mathbf{x}_t\right).$$

Its mean is the return of mean vector, and its variance is proportional to the projection of $\mathbf{x}_t$ on Σ.

According to the mean reversion idea, the probability of a profitable $\mathbf{b}$ with respect to a predefined mean reversion threshold ϵ is defined as

$$\mathrm{Pr}_{\mathbf{b} \sim \mathcal{N}(\mu, \Sigma)}[D \leq \epsilon] = \mathrm{Pr}_{\mathbf{b} \sim \mathcal{N}(\mu, \Sigma)}\left[\mathbf{b}^\top \mathbf{x}_t \leq \epsilon\right].$$

For simplicity, we write $\mathrm{Pr}[\mathbf{b}^\top \mathbf{x}_t \leq \epsilon]$ instead. Note that we are considering the mean reversion profitability in a portfolio consisting of multiple stocks; thus, this definition is equivalent to the motivating idea of buying poor-performing stocks or, equivalently, selling good-performing stocks.

The algorithm adjusts the distribution to ensure that the probability of a mean reversion profitable $\mathbf{b}$ is higher than a confidence-level parameter $\theta \in [0, 1]$:

$$\mathrm{Pr}\left[\mathbf{b}^\top \mathbf{x}_t \leq \epsilon\right] \geq \theta.$$

This is somewhat counterintuitive but reasonable with respect to the mean reversion idea. If it is highly probable that the portfolio return $\mathbf{b}^\top \mathbf{x}_t$ is less than a threshold, it is also highly probable that its next return based on $\mathbf{x}_{t+1}$ tends to be higher since $\mathbf{x}_{t+1}$ will revert.

Then, following the intuition underlying PA algorithms (Crammer et al. 2006), our algorithm chooses a distribution closest to the current distribution $\mathcal{N}(\mu_t, \Sigma_t)$ in terms of Kullback–Leibler (KL) divergence (Kullback and Leibler 1951). As a result, at the end of period t, the algorithm updates the distribution by solving the following optimization problem.

The Raw Optimization Problem: CWMR

$$(\mu_{t+1}, \Sigma_{t+1}) = \arg\min D_{\mathrm{KL}}(\mathcal{N}(\mu, \Sigma) \| \mathcal{N}(\mu_t, \Sigma_t))$$
$$\text{s.t. } \mathrm{Pr}[\mathbf{b}^\top \mathbf{x}_t \leq \epsilon] \geq \theta \qquad (10.1)$$
$$\mu \in \Delta_m.$$

The optimization problem (10.1) clearly reflects our motivation. On the one hand, if the current μ_t is mean reversion profitable, that is, the first constraint is satisfied, CWMR chooses the same distribution, resulting in a passive CRP strategy. On the other hand, if μ_t does not satisfy the mean reversion constraint, CWMR tries to figure out a new distribution, which is expected to profit and not far from the current distribution.

Let us reformulate the objective and constraints. For the objective part, the KL divergence between two Gaussian distributions can be rewritten as

$$D_{\mathrm{KL}}(\mathcal{N}(\mu, \Sigma) \| \mathcal{N}(\mu_t, \Sigma_t))$$
$$= \frac{1}{2}\left(\log\left(\frac{\det\Sigma_t}{\det\Sigma}\right) + \mathrm{Tr}(\Sigma_t^{-1}\Sigma) + (\mu_t - \mu)^\top \Sigma_t^{-1}(\mu_t - \mu) - d\right).$$

For the constraint part, since $\mathbf{b} \sim \mathcal{N}(\mu, \Sigma)$, $\mathbf{b}^\top \mathbf{x}_t$ has a univariate Gaussian distribution with mean $\mu_D = \mu^\top \mathbf{x}_t$ and variance $\sigma_D^2 = \mathbf{x}_t^\top \Sigma \mathbf{x}_t$. Thus, the probability of a return less than ϵ is

$$\Pr[D \leq \epsilon] = \Pr\left[\frac{D - \mu_D}{\sigma_D} \leq \frac{\epsilon - \mu_D}{\sigma_D}\right].$$

In the preceding equation, $\frac{D - \mu_D}{\sigma_D}$ is a normally distributed random variable; thus, the probability equals $\Phi\left(\frac{\epsilon - \mu_D}{\sigma_D}\right)$, where Φ is the cumulative distribution function of Gaussian distribution. As a result, we can rewrite the constraint as $\frac{\epsilon - \mu_D}{\sigma_D} \geq \Phi^{-1}(\theta)$. Substituting μ_D and σ_D by their definitions and rearranging the terms, we can obtain

$$\epsilon - \mu^\top \mathbf{x}_t \geq \phi \sqrt{\mathbf{x}_t^\top \Sigma \mathbf{x}_t},$$

where $\phi = \Phi^{-1}(\theta)$. Clearly, we require that the weighted summation of return and standard deviation is less than a threshold. Till now, we can rewrite the preceding optimization problem.

The Revised Optimization Problem: CWMR

$$(\mu_{t+1}, \Sigma_{t+1}) = \arg\min \ \frac{1}{2}\left(\log\left(\frac{\det\Sigma_t}{\det\Sigma}\right) + \mathrm{Tr}(\Sigma_t^{-1}\Sigma) + (\mu_t - \mu)^\top \Sigma_t^{-1}(\mu_t - \mu)\right)$$

$$\text{s.t. } \epsilon - \mu^\top \mathbf{x}_t \geq \phi \sqrt{\mathbf{x}_t^\top \Sigma \mathbf{x}_t}$$

$$\mu^\top \mathbf{1} = 1, \quad \mu \succeq 0. \tag{10.2}$$

For the optimization problem (10.2), the first constraint is not convex in Σ, therefore we have two ways to handle it. The first way (Dredze et al. 2008) is to linearize it by omitting the square root, that is, $\epsilon - \mu^\top \mathbf{x}_t \geq \phi \mathbf{x}_t^\top \Sigma \mathbf{x}_t$. As a result, we can finalize the first optimization problem, named CWMR-Var.

The Final Optimization Problem 1: CWMR-Var

$$(\mu_{t+1}, \Sigma_{t+1}) = \arg\min \ \frac{1}{2}\left(\log\left(\frac{\det\Sigma_t}{\det\Sigma}\right) + \mathrm{Tr}(\Sigma_t^{-1}\Sigma) + (\mu_t - \mu)^\top \Sigma_t^{-1}(\mu_t - \mu)\right)$$

$$\text{s.t. } \epsilon - \mu^\top \mathbf{x}_t \geq \phi \mathbf{x}_t^\top \Sigma \mathbf{x}_t$$

$$\mu^\top \mathbf{1} = 1, \quad \mu \succeq 0. \tag{10.3}$$

The second reformulation (Crammer et al. 2008) is to decompose the positive semidefinite (PSD) Σ, that is, $\Sigma = \Upsilon^2$ with $\Upsilon = Q\mathrm{diag}(\lambda_1^{1/2}, \ldots, \lambda_m^{1/2})Q^\top$, where Q is orthonormal and $\lambda_1, \ldots, \lambda_m$ are the eigenvalues of Σ and thus Υ is also PSD. This reformulation yields the second final optimization problem, named CWMR-Stdev.

The Final Optimization Problem 2: CWMR-Stdev

$$(\mu_{t+1}, \Upsilon_{t+1}) = \arg\min \frac{1}{2}\left(\log\left(\frac{\det\Upsilon_t^2}{\det\Upsilon^2}\right) + \mathrm{Tr}(\Upsilon_t^{-2}\Upsilon^2) + (\mu_t - \mu)^\top \Upsilon_t^{-2}(\mu_t - \mu)\right)$$

$$\text{s.t. } \epsilon - \mu^\top \mathbf{x}_t \geq \phi\|\Upsilon \mathbf{x}_t\|, \quad \Upsilon \text{ is PSD}$$

$$\mu^\top \mathbf{1} = 1, \quad \mu \succeq 0. \tag{10.4}$$

Clearly, the revised optimization problem (10.2) is equivalent to the raw optimization problem (10.1). From the revised problem, we proposed two final optimization problems, Equations 10.3 and 10.4, which are convex and thus can be efficiently solved by convex optimization (Boyd and Vandenberghe 2004). The first variation, CWMR-Var, linearizes the constraint; thus, it results in an *approximate* solution for the revised and the raw optimizations. In contrast, the second variation, CWMR-Stdev, is equivalent to the revised optimization problem (10.2) and results in an *exact* solution for both the revised and raw optimization problems.

Remarks on Formulations: Note that the short version of this chapter (Li et al. 2011b) assumes log utility (Bernoulli 1954; Latané 1959) on $\mu^\top \mathbf{x}_t$ and is slightly different from this version (Li et al. 2013). Since both ϵ and ϕ are adjustable, they have similar effects on μ. Assuming other parameters are constant except μ, as $\mu^\top \mathbf{x}_t > \log \mu^\top \mathbf{x}_t$, the current linear form can move μ toward the mean reversion profitable portfolio more than the log form can. However, the log form in this constraint causes another convexity issue besides the standard deviation on the right-hand side. To solve the optimization problem with log, Li et al. (2011b) chose to replace the log term by its linear approximation, which may converge to a different solution. Moreover, the current form and log's linear approximation are essentially the same.* Thus, we adopt return without log, which has no above convexity issues concerning log and its linear approximation.

10.3 Algorithms

Now, let us devise the proposed algorithms based on the optimization problem, Equations 10.3 and 10.4. Their solutions are shown in Propositions 10.1 and 10.2, respectively. Both proofs are presented in Appendices B.3.1 and B.3.2, respectively.

Proposition 10.1 The solution to the final optimization problem (10.3) (CWMR-Var) without considering the non-negativity constraint ($\mu \succeq 0$) is expressed as

$$\mu_{t+1} = \mu_t - \lambda_{t+1}\Sigma_t(\mathbf{x}_t - \bar{x}_t \mathbf{1}), \quad \Sigma_{t+1}^{-1} = \Sigma_t^{-1} + 2\lambda_{t+1}\phi\mathbf{x}_t\mathbf{x}_t^\top,$$

where λ_{t+1} corresponds to the Lagrangian multiplier calculated as Equation B.10 in Appendix B.3.1 and $\bar{x}_t = \frac{\mathbf{1}^\top \Sigma_t \mathbf{x}_t}{\mathbf{1}^\top \Sigma_t \mathbf{1}}$ denotes the CW average of $\mathbf{x}_t$.

---

*Current form is $\mu^\top \mathbf{x}_t$. Li et al. (2011b) uses approximation, that is, $\log \mu^\top \mathbf{x}_t \approx \log \mu_t^\top \mathbf{x}_t + \frac{\mathbf{x}_t \cdot (\mu - \mu_t)}{\mu_t^\top \mathbf{x}_t}$. Both terms are linear, although their scales are different.

Proposition 10.2 The solution to the final optimization problem (10.4) (CWMR-Stdev) without considering the non-negativity constraint ($\mu \succeq 0$) is expressed as

$$\mu_{t+1} = \mu_t - \lambda_{t+1}\Sigma_t(\mathbf{x}_t - \bar{x}_t\mathbf{1}), \quad \Sigma_{t+1}^{-1} = \Sigma_t^{-1} + \lambda_{t+1}\phi\frac{\mathbf{x}_t\mathbf{x}_t^\top}{\sqrt{U_t}},$$

where λ_{t+1} denotes the Lagrangian multiplier calculated as Equation B.14 in Appendix B.3.2, $\bar{x}_t = \frac{\mathbf{1}^\top\Sigma_t\mathbf{x}_t}{\mathbf{1}^\top\Sigma_t\mathbf{1}}$ represents the CW average of $\mathbf{x}_t$, and $V_t = \mathbf{x}_t^\top\Sigma_t\mathbf{x}_t$ and $\sqrt{U_t} = \frac{-\lambda_{t+1}\phi V_t + \sqrt{\lambda_{t+1}^2\phi^2 V_t^2 + 4V_t}}{2}$ denote the return variances for period t and $t+1$, respectively.

Initially, with no information available for the task, we simply initialize μ_1 to uniform and each diagonal element of the covariance matrix Σ_1 to $\frac{1}{m^2}$, or equivalent standard deviation $\frac{1}{m}$. Note that we solve the optimization problems by ignoring the non-negativity constraint ($\mu \succeq 0$), which is a typical way to reduce the complexity (Helmbold et al. 1998; Agarwal et al. 2006). To solve this issue that μ can be negative, we simply project the resulting μ to the simplex domain (Agarwal et al. 2006). In the context of investment, this means that we firstly allow shorting, and later lower the leverage* by a simplex projection. Another remaining issue is that, although the covariance matrix is nonsingular in theory, in real computation, Σ sometimes may be singular due to computer precision. To avoid this problem and be consistent with the projection of μ, we rescale Σ by normalizing its summation value to $\frac{1}{m}$, which equals the sum of elements in μ_1. Note that we arbitrarily choose $\frac{1}{m}$, while one can choose other values, which generally do not affect the performance too much. The final CWMR algorithms are presented in Algorithm 10.1, and OLPS with both deterministic and stochastic CWMR algorithms is illustrated in Algorithm 10.2.

The algorithms have two possible parameters, that is, confidence parameter ϕ and mean reversion parameter ϵ. Typically, the first parameter, ϕ, can be 1.28, 1.64, 1.95, or 2.57, with corresponding θ values of 80%, 90%, 95%, or 99%. As we have tested, ϕ does not affect the final performance too much. On the contrary, ϵ has a significant impact on the final performance. As our model is long-only,[†] we put more weights on the poor-performing assets; thus, ϵ is often in the range of [0, 1]. On the one hand, if the value is too large, such as $\epsilon \geq 1.2$, the last portfolio distribution can always satisfy the constraint and requires no updates. In such a situation, with an initial uniform portfolio, CWMR will degrade to uniform CRP. On the other hand, if the value is too small, such as $\epsilon \leq 0.5$, the last distribution can always dissatisfy the constraint and has to be frequently updated. In between, CWMR updates the distribution when the last distribution cannot satisfy the constraint. We will further validate the above analysis by evaluating its parameter effect in Section 13.3.3.

---

*In investment, notional leverage denotes total holding assets plus total notional amount of liability divided by equity. If shorting is allowed, the notional leverage equals $\sum_i |b_i|$ to 1. The problem setting in our study is long-only, in which we do not allow shoring/margin; thus, the leverage is always 1 to 1.

†Long-only means no shoring/margin is allowed; thus, the notional leverage is always 1 to 1.

Algorithm 10.1: Confidence-Weighted Mean Reversion:
CWMR(ϕ, ϵ, (μ_t, Σ_t), $\mathbf{x}_1^t$, t).

Input: ϕ: Confidence parameter; $\epsilon \in [0, 1]$: Mean reversion parameter;
(μ_t, Σ_t): Current portfolio distribution; $\mathbf{x}_1^t$: Historical market
sequence; t: Index of current trading period.
Output: $(\mu_{t+1}, \Sigma_{t+1})$: Next portfolio distribution.
begin

Calculate the following variables:

$$M_t = \mu_t^\top \mathbf{x}_t, \quad V_t = \mathbf{x}_t^\top \Sigma_t \mathbf{x}_t, \quad W_t = \mathbf{x}_t^\top \Sigma_t \mathbf{1}, \quad \bar{x}_t = \frac{\mathbf{1}^\top \Sigma_t \mathbf{x}_t}{\mathbf{1}^\top \Sigma_t \mathbf{1}}$$

Update the portfolio distribution:

$$\text{CWMR-Var} \begin{cases} \lambda_{t+1} \text{ as in Equation B.10 in Appendix B.3.1} \\ \mu_{t+1} = \mu_t - \lambda_{t+1} \Sigma_t (\mathbf{x}_t - \bar{x}_t \mathbf{1}) \\ \Sigma_{t+1} = (\Sigma_t^{-1} + 2\lambda_{t+1} \phi \, \text{diag}^2(\mathbf{x}_t))^{-1} \end{cases}$$

$$\text{CWMR-Stdev} \begin{cases} \lambda_{t+1} \text{ as in Equation B.14 in Appendix B.3.2} \\ \sqrt{U_t} = \frac{-\lambda_{t+1}\phi V_t + \sqrt{\lambda_{t+1}^2 \phi^2 V_t^2 + 4V_t}}{2} \\ \mu_{t+1} = \mu_t - \lambda_{t+1} \Sigma_t (\mathbf{x}_t - \bar{x}_t \mathbf{1}) \\ \Sigma_{t+1} = (\Sigma_t^{-1} + \lambda_{t+1} \frac{\phi}{\sqrt{U_t}} \text{diag}^2(\mathbf{x}_t))^{-1} \end{cases}$$

Normalize μ_{t+1} and Σ_{t+1}:

$$\mu_{t+1} = \arg\min_{\mu \in \Delta_m} \|\mu - \mu_{t+1}\|^2, \quad \Sigma_{t+1} = \frac{\Sigma_{t+1}}{m \text{Tr}(\Sigma_{t+1})}$$

end

10.4 Analysis

In this section, we analyze and interpret the proposed algorithms. Firstly, we compare CWMR with CW learning (Crammer et al. 2008; Dredze et al. 2008). Then, we analyze CWMR's update schemes, that is, μ and Σ, with running examples. Further, we describe the behavior of stochastic CWMR. Finally, we show its computational time and compare it with existing work.

The proposed CWMR algorithms are partially motivated by CW learning, thus their formulations and subsequent derivations are similar. However, they address different problems, as CWMR handles OLPS while CW focuses on classification. Although both objectives adopt KL divergence to measure the closeness between two distributions, their constraints reflect that they are oriented toward different problems. To be specific, CW's constraint is the probability of a correct classification, while

---

Algorithm 10.2: Online portfolio selection with CWMR.

Input: $\phi = \Phi^{-1}(\theta)$: Confidence parameter; $\epsilon \in [0, 1]$: Mean reversion
parameter; $\mathbf{x}_1^n$: Historical market sequence.
Output: S_n: Final cumulative wealth.
begin

  Initialization: $t = 1$, $\mu_1 = \frac{1}{m}\mathbf{1}$, $\Sigma_1 = \frac{1}{m^2}\mathbf{I}$, $S_0 = 1$;

  for $t = 1, \ldots, n$ **do**

    Draw a portfolio $\mathbf{b}_t$ from $\mathcal{N}(\mu_t, \Sigma_t)$:

    Deterministic CWMR : $\quad \mathbf{b}_t = \mu_t$

    Stochastic CWMR : $\quad \tilde{\mathbf{b}}_t \sim \mathcal{N}(\mu_t, \Sigma_t), \quad \mathbf{b}_t = \underset{\mathbf{b} \in \Delta_m}{\arg\min} \|\mathbf{b} - \tilde{\mathbf{b}}_t\|^2$

    Receive stock price relatives: $\mathbf{x}_t = (x_{t1}, \ldots, x_{tm})$;
    Calculate the daily return and cumulative return: $S_t = S_{t-1} \times (\mathbf{b}_t^\top \mathbf{x}_t)$;
    Update the portfolio distribution:
    $(\mu_{t+1}, \Sigma_{t+1}) = \text{CWMR}(\phi, \epsilon, (\mu_t, \Sigma_t), \mathbf{x}_1^t, t)$;

  end

end

CWMR's constraints are the probability of an underperforming portfolio plus the
simplex constraint. If there is mean reversion, the portfolio should be a profitable
one, in the next period. Their formulations' differences result in subsequent different
derivations.

Then, we provide a preliminary analysis on μ, which is the main concern for
CWMR, to reflect its underlying mean reversion idea. Both CWMR-Var and CWMR-
Stdev share the same update on μ, that is,

$$\mu_{t+1} = \mu_t - \lambda_{t+1}\Sigma_t(\mathbf{x}_t - \bar{x}_t\mathbf{1}).$$

Straightforwardly, we can rewrite its term as $\mu_{t+1,i} = \mu_{t,i} - \lambda_{t+1}\sigma_t^2(x_{t,i} - \bar{x}_t)$. Obviously, λ_{t+1} is non-negative and Σ_t is PSD. The term $\mathbf{x}_t - \bar{x}_t\mathbf{1}$ denotes excess return
vector for period t, where $\bar{x}_t$ is the CW average of $\mathbf{x}_t$. Holding other terms constant,
μ_{t+1} tends to move toward μ_t, while the magnitude is negatively related to the last
excess return, which is the mean reversion principle. Meanwhile, these movements
are dynamically adjusted by λ_{t+1}, the last covariance matrix Σ_t and mean μ_t, which
catch both first- and second-order information. To the best of our knowledge, none
of the existing algorithms has explicitly exploited the second-order information of $\mathbf{b}$,
even though the second-order information could benefit the proposed algorithms.

Let us continue to analyze Σ. With only nonzero diagonal elements, we can write
the update of the i-th variance as

$$\sigma^2 = \sigma_i^2/(1 + \lambda_{t+1}\phi' x_{ti}^2\sigma_i^2),$$

where $\phi' = 2\phi$ for CWMR-Var and $\phi' = \frac{\phi}{\sqrt{U_t}}$ for CWMR-Stdev. Since both λ_{t+1} and ϕ' are positive, poor-performing assets (with lower values of $x_{t,i}$) have higher variance terms than good-performing ones (with higher $x_{t,i}$). Note that Σ denotes the covariance matrix of $\mathbf{b}$ rather than $\mathbf{x}$. Thus, a higher value means that the corresponding mean is more volatile than others. Since we move the weights from good-performing assets to poor-performing ones, the latter would change more than the former, that is, the latter has higher volatility. In the next update of μ, assets with high volatility would actively magnify the movement magnitude.

To better illustrate the updates, we give running updates based on a classic example (Cover and Gluss 1986). Let a market consist of cash and one volatility asset, and the sequence of $\mathbf{x}$ is $\left(1, \frac{1}{2}\right)$, $(1, 2)$, $\left(1, \frac{1}{2}\right)$, $\ldots$. Obviously, market strategy can gain nothing since no asset grows in the long run. The best CRP strategy, with $\mathbf{b} = \left(\frac{1}{2}, \frac{1}{2}\right)$, grows to $\left(\frac{9}{8}\right)^{\frac{n}{2}}$ at the end of n periods. However, starting with $\mu_0 = \left(\frac{1}{2}, \frac{1}{2}\right)$, the CWMR strategy can grow to $\frac{3}{4} \times 2^{\frac{n-1}{2}}$ after n-th periods. Table 10.2 shows the running details for the initial five periods, and further details can be derived. On each period $t + 1$, the mean moves toward the last mean and also moves far away by the excess return vector $(\mathbf{x}_t - \bar{x}_t \mathbf{1})$, and its magnitude is determined by both λ_t and Σ_t. Note that, in this example, μ before projection is out of the simplex domain and is forced sparse via normalization, which is not an usual case in real tests. In summary, both the first- and second-order information contribute to CWMR's success.

Then, let us compare deterministic CWMR with the stochastic version (Line 4 in Algorithm 10.2), which draws a portfolio based on both the mean and covariance matrix. Interestingly, Σ negatively affects CWMR's performance in several aspects. Firstly, a stochastic $\mathbf{b}$ drawn from the distribution is always different from the optimal mean μ, which obviously causes performance divergences. Given that Σ converges to the zero matrix (see the recursive updates in the two propositions), the distribution of $\mathbf{b}$ conditioning on the data converges to the point mass at the mean parameter value, $\mu = \lim_t \mu_t$. Thus, drawing weights $\mathbf{b}$ from the distribution (the stochastic version) is suboptimal, since we already have an estimate of μ. It is better to choose $\mathbf{b}$ as either the mode or mean (incidentally, the same for the Gaussian case), which is actually deterministic CWMR. Another effect caused by the stochastic behavior is

Table 10.2 *A running example of CWMR-Stdev on the Cover's game*

t	$\mathbf{x}_t$	$\mathbf{b}_t$	$\mathbf{b}_t^\top \mathbf{x}_t$	λ_t	$\mathbf{x}_t - \bar{x}_t \mathbf{1}$	$\mathrm{diag}(\Sigma_t)$	μ_t
0						$(0.25, 0.25)$	$(0.5, 0.5)$
1	$(1.0, 0.5)$	$(0.5, 0.5)$	0.75	40.78	$(0.25, -0.25)$	$(0.10, 0.40)$	$(0.0, 1.0)$
2	$(1.0, 2.0)$	$(0.0, 1.0)$	2.00	61.61	$(-0.80, 0.20)$	$(0.40, 0.10)$	$(1.0, 0.0)$
3	$(1.0, 0.5)$	$(1.0, 0.0)$	1.00	75.56	$(0.10, -0.40)$	$(0.10, 0.40)$	$(0.0, 1.0)$
4	$(1.0, 2.0)$	$(0.0, 1.0)$	2.00	61.61	$(-0.80, 0.20)$	$(0.40, 0.10)$	$(1.0, 0.0)$
5	$(1.0, 0.5)$	$(1.0, 0.0)$	1.00	75.56	$(0.10, -0.40)$	$(0.10, 0.40)$	$(0.0, 1.0)$
$\vdots$	$\vdots$	$\vdots$	$\vdots$	$\vdots$	$\vdots$	$\vdots$	$\vdots$

Table 10.3 *Summary of time complexity analysis*

Methods	Time Complexity	Methods	Time Complexity
UP	$O(n^m)/O(m^7n^8)$	SP/GRW/M0	$O(mn)$
EG	$O(mn)$	Anticor	$O(N^3m^2n)$
ONS	$O(m^3n)$	B^K/B^{NN}/CORN	$O(N^2mn^2)+O(Nmn^2)$
PAMR	$O(mn)$	CWMR	$O(mn)$

the additional projection, as sometimes the stochastic **b** may be out of the simplex domain. To better understand the two aspects, let us continue the Cover's game in Table 10.2. For the first case, let $\mu = (0.5, 0.5)$ and $\text{diag}(\mathbf{\Sigma}) = (0.25, 0.25)$. We draw stochastic **b** for 10,000 times, and the average **b** after projection is $(0.5038, 0.4962)$ (before projection, the value is $(0.5070, 0.4993)$), which slightly deviates from the optimal mean and will result in different performance. For the second case, let $\mu = (0, 1)$ and $\text{diag}(\mathbf{\Sigma}) = (0.1, 0.4)$. We draw and project 10,000 stochastic **b**, and get an average **b** of $(0.1391, 0.8609)$, which is far from the optimal mean $(0, 1)$. In both cases, stochastic CWMR tends to deviate from the optimal mean, and thus underperforms the deterministic one, which is shown in the related experiments (Li et al. 2013, Table VII).

Since computational time is of crucial importance for certain trading scenarios, such as high-frequency trading (Aldridge 2010), which can occur in fractions of a second, we finally show CWMR's time complexity. In the implementation, we only consider the diagonal elements of $\mathbf{\Sigma}$; thus, its inverse costs linear time. Moreover, the projection (Line 3 in Algorithm 10.1) can be implemented in $O(m)$ time (Duchi et al. 2008). Thus, in total, CWMR algorithms (Algorithm 10.1) take $O(m)$ time per period. Straightforwardly, OLPS with CWMR (Algorithm 10.2) takes $O(mn)$ time. Table 10.3 compares CWMR's time complexity with that of existing strategies.* Clearly, CWMR takes no more time than any others.

10.5 Summary

In this chapter, we proposed a novel online portfolio selection (OLPS) strategy named confidence-weighted mean reversion (CWMR), which effectively learns portfolios by exploiting the mean reversion property in financial markets and the second-order information of a portfolio. CWMR's update schemes are obtained by solving two optimization problems that consider both first- and second-order information of a portfolio vector, which goes beyond any existing approaches that only consider first-order information. As shown in Part IV, the proposed approach beats a number of

---

*Nonparametric learning approaches (B^K, B^{NN}, and CORN) require to solve a nonlinear optimization each period, that is, $\mathbf{b}_{t+1} = \arg\max_{\mathbf{b}\in\Delta_m} \prod_i (\mathbf{b}^\top \mathbf{x}_i)$, whose time complexity is generally high. To produce an approximate solution, batch gradient projection algorithms (Helmbold et al. 1997) take $O(mn)$, while the batch convex Newton method (Agarwal et al. 2006) takes $O(m^3n)$. In the table, we set the step $O(mn)$ time complexity. In our implementation, we adopt MATLAB® Optimization Toolbox™ (function *fmincon* with *active-set*) to obtain exact solutions.

competing state-of-the-art approaches on various up-to-date datasets collected from the real market.

In future, we plan to study in detail the cause behind the existence of the mean reversion property in the financial markets. This will help us to further understand the nature of the markets. Second, we also intend to explore the possibility of combining both the trend-following and mean reversion principles to provide more practically effective solutions. Finally, we note that an interesting future direction is to extend our analysis for long-short portfolios.*

---

*Long-short portfolios can have negative weights, which denote the short positions.

Chapter 11

Online Moving Average Reversion

Empirical evidence shows that a stock's high and low prices are temporary, and stock price relatives are likely to follow the mean reversion phenomenon. While existing mean reversion strategies can achieve good empirical performance on many real datasets, they often make a *single-period mean reversion* assumption, which is not always satisfied, leading to poor performance on some real datasets. To overcome the limitation, this chapter (Li et al. 2015) proposes a *multiple-period mean reversion*, or the so-called moving average reversion (MAR), and a new online portfolio selection (OLPS) strategy named the online moving average reversion (OLMAR), which exploits MAR by applying powerful online learning techniques. Our empirical evaluations in Part IV show that OLMAR can overcome the drawbacks of existing mean reversion algorithms and achieve significantly better results, especially on the datasets where existing mean reversion algorithms failed. In addition to superior performance, OLMAR also runs extremely fast, further supporting its practical applicability to a wide range of applications.

This chapter is organized as follows. Section 11.1 analyzes existing works and motivates the proposed strategy. Section 11.2 formulates the strategy, and Section 11.3 solves the formulations and derives the algorithms. Section 11.4 further analyzes the proposed algorithm. Finally, Section 11.5 summarizes this chapter and indicates future directions.

11.1 Preliminaries

11.1.1 Related Work

Most existing formulations follow the basic routine of Kelly-based portfolio selection (Kelly 1956; Thorp 1971). In particular, a portfolio manager predicts $\tilde{\mathbf{x}}_{t+1}$ in terms of k possible values $\tilde{\mathbf{x}}_{t+1}^1, \ldots, \tilde{\mathbf{x}}_{t+1}^k$ and their corresponding probabilities $p_1, \ldots, p_k$. Note that each $\tilde{\mathbf{x}}_{t+1}^i$ denotes one possible combination vector of individual price relative predictions. Then, he or she can figure out a portfolio by maximizing the expected log return,

$$\mathbf{b}_{t+1} = \arg\max_{\mathbf{b} \in \Delta_m} \sum_{i=1}^{k} p_i \log(\mathbf{b} \cdot \tilde{\mathbf{x}}_{t+1}^i).$$

Based on the methods to predict $\tilde{\mathbf{x}}_{t+1}^i$ and p_i, most existing algorithms can be classified into three categories. Table 11.1 summarizes their optimization formulations and underlying prediction schemes, whose details can be found on their respective studies. Note that we have transformed certain formulations without changing their key ideas.

Now let us introduce the three categories according to their empirical performance. The second category, which consists of successive constant rebalanced portfolio (SCRP) and online Newton step (ONS), assumes that the predictions consist of all historical price relatives with uniform distribution. That is, at period $t + 1$, the price relative vector may be $\mathbf{x}_i$, $i = 1, \ldots, t$ with a probability of $\frac{1}{t}$. In other words, this category aims to model the next price relatives as their historical average. The algorithms in this category all present good theoretical regret bound and are universal. However, their empirical results show that such an assumption may be inappropriate to model the market behaviors. The third category, which mainly consists of the pattern matching–based algorithms, models the next price relatives as a sampled set of similar price relatives. In particular, denoting the similar index set as C_t, it models the next price relative vector as $\mathbf{x}_i$, $i \in C_t$, with a uniform probability of $\frac{1}{|C_t|}$. The algorithms in this category (except correlation-driven nonparametric learning [CORN]) enjoy the universal consistency property, and their empirical results also show that such an assumption can explain the markets well.

Algorithms in the first category, which consists of exponential gradient (EG), passive–aggressive mean reversion (PAMR), and confidence-weighted mean reversion (CWMR), assume a single prediction value with a probability of 100% and maintain previous portfolio information via regularization techniques. In particular, EG assumes $\tilde{\mathbf{x}}_{t+1}^1 = \mathbf{x}_t$ with $p_1 = 100\%$, while PAMR and CWMR assume $\tilde{\mathbf{x}}_{t+1}^1 = \frac{1}{\mathbf{x}_t}$* with $p_1 = 100\%$, which is in essence mean reversion. Note that the formulations of PAMR and CWMR ignore the log utility due to the single-value prediction and the consideration of convexity and computation. Though all three algorithms assume that all information is fully reflected by $\mathbf{x}_t$, their performance diverges and supports that mean reversion may better explain the markets. On the one hand, even with a decent theoretical result, EG always performs poorly. On the other hand, though without theoretical guarantees, PAMR and CWMR have produced the best results in certain real markets. However, when such a single-period mean reversion assumption is not satisfied, PAMR and CWMR would suffer from dramatic failures (Li et al. 2012, Table 4, the DJIA dataset), which motivates the following approach.

11.1.2 Motivation

Empirical results (Li et al. 2011b, 2012) show that mean reversion, which assumes the poor stock may perform well in the subsequent periods, may better explain the markets. PAMR and CWMR can exploit the mean reversion property well and achieve good results on most datasets at the time, especially on the New York Stock Exchange benchmark dataset (Cover 1991). However, they rely on a naïve assumption that next

---

*This assumption requires some transformations. That is, given $\mathbf{x}_t \in \mathbb{R}_m^+$, minimizing $\mathbf{b} \cdot \mathbf{x}_t$ is equivalent to maximizing $\mathbf{b} \cdot \frac{1}{\mathbf{x}_t}$. The latter follows the analysis framework here.

Table 11.1 *Summary of existing optimization formulations and their underlying predictions*

Categories	Methods	Formulations	Prediction ($\tilde{\mathbf{x}}_{t+1}^i$)	Probability (p_i)
In hindsight	BCRP	$\mathbf{b}_{t+1} = \arg\max_{\mathbf{b}\in\Delta_m} \sum_{i=1}^n \frac{1}{n}\log\mathbf{b}\cdot\mathbf{x}_i$	$\mathbf{x}_i, i = 1,\ldots,n$	$1/n$
1	EG	$\mathbf{b}_{t+1} = \arg\max_{\mathbf{b}\in\Delta_m}\log\mathbf{b}\cdot\mathbf{x}_t - \lambda R(\mathbf{b},\mathbf{b}_t)$	$\mathbf{x}_t$	1.00
	PAMR	$\mathbf{b}_{t+1} = \arg\min_{\mathbf{b}\in\Delta_m}\mathbf{b}\cdot\mathbf{x}_t + \lambda R(\mathbf{b},\mathbf{b}_t)$	$1/\mathbf{x}_t$	1.00
	CWMR	$\mathbf{b}_{t+1} = \arg\min_{\mathbf{b}\in\Delta_m}\text{Prob}(\mathbf{b}\cdot\mathbf{x}_t) + \lambda R(\mathbf{b},\mathbf{b}_t)$	$1/\mathbf{x}_t$	1.00
2	SCRP	$\mathbf{b}_{t+1} = \arg\max_{\mathbf{b}\in\Delta_m}\sum_{i=1}^t \frac{1}{t}\log\mathbf{b}\cdot\mathbf{x}_i$	$\mathbf{x}_i, i = 1,\ldots,t$	$1/t$
	ONS	$\mathbf{b}_{t+1} = \arg\max_{\mathbf{b}\in\Delta_m}\sum_{i=1}^t \frac{1}{t}\log\mathbf{b}\cdot\mathbf{x}_i - \lambda R(\mathbf{b})$	$\mathbf{x}_i, i = 1,\ldots,t$	$1/t$
3	B^K/B^{NN}/CORN	$\mathbf{b}_{t+1} = \arg\max_{\mathbf{b}\in\Delta_m}\sum_{i\in C_t}\frac{1}{|C_t|}\log\mathbf{b}\cdot\mathbf{x}_i$	$\mathbf{x}_i, i \in C_t$	$1/|C_t|$

Note: $R(\cdot)$ and $R(\cdot,\cdot)$ denote regularization terms, such as L_2 norm. PAMR/CWMR's prediction is not of strictly equivalence, which we do not prove.

price relative $\tilde{\mathbf{x}}_{t+1}$ will be inversely proportional to last price relative $\mathbf{x}_t$. In particular, they implicitly assume that next price $\tilde{\mathbf{p}}_{t+1}$ will revert to last price $\mathbf{p}_{t-1}$,

$$\tilde{\mathbf{x}}_{t+1} = \frac{1}{\mathbf{x}_t} \implies \frac{\tilde{\mathbf{p}}_{t+1}}{\mathbf{p}_t} = \frac{\mathbf{p}_{t-1}}{\mathbf{p}_t} \implies \tilde{\mathbf{p}}_{t+1} = \mathbf{p}_{t-1}.$$

Note that both $\mathbf{x}$ and $\mathbf{p}$ are vectors and the above operations are element-wise.

Though empirically effective on most datasets, PAMR and CWMR's *single-period* assumption causes two potential problems. Firstly, both algorithms suffer from frequently fluctuating raw prices, as they often contain a lot of noise. Secondly, their assumption of single-period mean reversion may not always be satisfied in the real world. Even two consecutive declining price relatives, which are common, can deactivate or fail both algorithms. One real example (Li et al. 2012) is the DJIA dataset (Borodin et al. 2004), on which PAMR performs the worst among the state of the art. Thus, traders are more likely to predict prices using some long-term values. Also on the DJIA dataset, Anticor, which exploits the *multiperiod* statistical correlation, performs much better than others. However, due to its heuristic nature (Li et al. 2011b, 2012), Anticor cannot fully exploit the mean reversion property. The two problems caused by the single-period assumption and Anticor's inability to fully exploit mean reversion call for a more powerful approach to effectively exploit mean reversion, especially in terms of multiple periods.

Now let us see a classic example (Cover and Gluss 1986) to illustrate the drawbacks of single-period mean reversion, as shown in Table 11.2. The toy market consists of cash and one volatile stock, whose market sequence follows A. It is easy to prove that best constant rebalanced portfolio (BCRP) ($\mathbf{b} = (\frac{1}{2}, \frac{1}{2})$) can grow by a factor of $(\frac{9}{8})^{n/2}$, while PAMR can grow by a better factor of $\frac{3}{2} \times 2^{(n-1)/2}$. Note that this virtual sequence is essentially single-period mean reversion, which perfectly fits with PAMR and CWMR's assumption. However, if market sequence does not satisfy such an assumption, both PAMR and CWMR would fail badly. Let us extend the market sequence to a two-period reversion, that is, market sequence B. In such a market, BCRP can achieve the same growth as before. Contrarily, PAMR can achieve a constant wealth $\frac{3}{2}$, which has no growth! More generally, if we further extend to k-period mean reversion, BCRP can still achieve the same growth, while PAMR will grow to $\frac{3}{2} \times (\frac{1}{2})^{(n-1) \times (\frac{1}{2} - \frac{1}{k})}$, which definitely approaches bankruptcy if $k \geq 3$.

To better exploit the (multiperiod) mean reversion property, we proposed a new type of algorithms, OLMAR, for OLPS. The essential idea is to exploit multiperiod moving average (mean) reversion via power online machine learning. Rather than $\tilde{\mathbf{p}}_{t+1} = \mathbf{p}_{t-1}$, OLMAR assumes that the next price will revert to a *moving average* (MA), that is, $\tilde{\mathbf{p}}_{t+1} = MA_t$, where MA_t denotes the MA till the end of period t. In time-series analysis, MA focuses on long-term trends and is typically used to smooth short-term price fluctuations, and thus can solve the two drawbacks of existing mean reversion algorithms.

Table 11.2 Illustration of the mean reversion strategies on toy markets

Market Sequences	BCRP	PAMR	OLMAR
A : $(1,2), (1,\frac{1}{2}), (1,2), (1,\frac{1}{2}), \ldots$	$(\frac{9}{8})^{n/2}$	$\frac{3}{2} \times 2^{\frac{n-1}{2}}$	$\frac{9}{8}$
B : $(1,2), (1,2), (1,\frac{1}{2}), (1,\frac{1}{2}), (1,2), \ldots$	$(\frac{9}{8})^{n/2}$	$\frac{3}{2}$	$\frac{9}{16} \times 2^{\frac{n-4}{2}}$
C : $(1,2), (1,2), (1,\frac{1}{2}), (1,\frac{1}{2}), (1,\frac{1}{2}), (1,2), \ldots$	$(\frac{9}{8})^{n/2}$	$\frac{3}{2} \times (\frac{1}{2})^{\frac{n-1}{6}}$	$\frac{9}{8} \times 2^{\frac{n-5}{6}}$
D : $\underbrace{(1,2), \ldots, (1,2)}_{k=4}, \underbrace{(1,1/2), \ldots, (1,1/2)}_{k=4}, (1,2), \ldots$	$(\frac{9}{8})^{n/2}$	$\frac{3}{2} \times (\frac{1}{2})^{\frac{n-1}{4}}$	$\frac{9}{4} \times 2^{\frac{n-6}{8}}$
E : $\underbrace{(1,2), \ldots, (1,2)}_{k=5}, \underbrace{(1,1/2), \ldots, (1,1/2)}_{k=5}, (1,2), \ldots$	$(\frac{9}{8})^{n/2}$	$\frac{3}{2} \times (\frac{1}{2})^{(n-1)\times\frac{3}{10}}$	$\frac{9}{2} \times 2^{\frac{n-7}{10}}$

Note: Since OLMAR is sensitive to the windows size, we set its windows to k. We calculate all OLMAR values with a mean reversion threshold of 2.

Without detailing the calculation,* we list the growth of OLMAR in different toy markets in Table 11.2. Clearly, OLMAR performs much better than PAMR in multiperiod mean reversion, but PAMR performs better than OLMAR in single-period reversion. Further empirical evaluations in Part IV show that the markets are more likely to follow multiperiod reversion.

11.2   Formulations

In this chapter, we adopt two types of moving average. The first, the so-called simple moving average (SMA), truncates the historical prices via a window and calculates its arithmetical average:

$$SMA_t(w) = \frac{1}{w} \sum_{i=t-w+1}^{t} \mathbf{p}_i,$$

where w denotes the window size and the summation is element-wise. Although we can enlarge the window size such that SMA can include more historical price relatives, the empirical evaluations in Part IV show that as the window size increases, its performance drops.

To consider entire price relatives rather than a window, the second type, exponential moving average (EMA), adopts all historical prices, and each price is exponentially weighted,

$$EMA_1(\alpha) = \mathbf{p}_1$$
$$EMA_t(\alpha) = \alpha\mathbf{p}_t + (1-\alpha)EMA_{t-1}(\alpha)$$
$$= \alpha\mathbf{p}_t + (1-\alpha)\alpha\mathbf{p}_{t-1} + (1-\alpha)^2\alpha\mathbf{p}_{t-2} + \cdots + (1-\alpha)^{t-1}\mathbf{p}_1,$$

where $\alpha \in (0, 1)$ denotes a decaying factor.

To this end, we can calculate the predicted price relative vector following the idea of the so-called moving average reversion (MAR). Based on the two types of moving average, we can infer two types of MAR.

Moving Average Reversion: MAR-1

$$\tilde{\mathbf{x}}_{t+1}(w) = \frac{SMA_t(w)}{\mathbf{p}_t} = \frac{1}{w}\left(\frac{\mathbf{p}_t}{\mathbf{p}_t} + \frac{\mathbf{p}_{t-1}}{\mathbf{p}_t} + \cdots + \frac{\mathbf{p}_{t-w+1}}{\mathbf{p}_t}\right)$$
$$= \frac{1}{w}\left(1 + \frac{1}{\mathbf{x}_t} + \cdots + \frac{1}{\bigodot_{i=0}^{w-2}\mathbf{x}_{t-i}}\right), \tag{11.1}$$

where w is the window size and $\bigodot$ denotes the element-wise product.

---

*We calculate OLMAR's growth using Algorithm 11.1. As the market sequences repeat themselves, OLMAR will finally stabilize.

Moving Average Reversion: MAR-2

$$\tilde{\mathbf{x}}_{t+1}(\alpha) = \frac{EMA_t(\alpha)}{\mathbf{p}_t} = \frac{\alpha\mathbf{p}_t + (1-\alpha)EMA_{t-1}(\alpha)}{\mathbf{p}_t}$$

$$= \alpha\mathbf{1} + (1-\alpha)\frac{EMA_{t-1}(\alpha)}{\mathbf{p}_{t-1}}\frac{\mathbf{p}_{t-1}}{\mathbf{p}_t} \qquad (11.2)$$

$$= \alpha\mathbf{1} + (1-\alpha)\frac{\tilde{\mathbf{x}}_t}{\mathbf{x}_t},$$

where $\alpha \in (0, 1)$ denotes the decaying factor and the operations are all element-wise.

Based on the expected price relative vector in Equations 11.1 and 11.2, OLMAR further adopts the idea of an effective online learning algorithm, that is, passive–aggressive (PA) (Crammer et al. 2006) learning, to exploit the MAR. Generally proposed for classification, PA passively keeps the previous solution if the classification is correct, while aggressively approaches a new solution if the classification is incorrect. After formulating the proposed OLMAR, we solve its closed-form update and design specific algorithms.

The proposed formulation, OLMAR, is to exploit MAR via PA online learning. The basic idea is to maximize the expected return $\mathbf{b} \cdot \tilde{\mathbf{x}}_{t+1}$ and keep last portfolio information via a regularization term. Thus, we follow the similar idea of PAMR (Li et al. 2012) and formulate an optimization as follows.

Optimization Problem: OLMAR

$$\mathbf{b}_{t+1} = \arg\min_{\mathbf{b}\in\Delta_m} \frac{1}{2}\|\mathbf{b} - \mathbf{b}_t\|^2 \quad \text{s. t.} \quad \mathbf{b} \cdot \tilde{\mathbf{x}}_{t+1} \geq \epsilon.$$

Note that we adopt expected return rather than expected log return. According to Helmbold et al. (1998), to solve the optimization with expected log return, one can adopt the first-order Taylor expansion, which is essentially linear. Such discussions are illustrated in Sections 9.2 and 10.2.

The above formulation explicitly reflects the basic idea of the proposed OLMAR. On the one hand, if its constraint is satisfied, that is, the expected return is higher than a threshold, then the resulting portfolio becomes equal to the previous portfolio. On the other hand, if the constraint is not satisfied, then the formulation will figure out a new portfolio such that the expected return is higher than the threshold, while the new portfolio is not far from the last one.

Since OLMAR follows the same learning principle as PAMR, their formulations are similar. However, the two formulations are essentially different. In particular, PAMR's core constraint (i.e., $\mathbf{b} \cdot \mathbf{x}_t \leq \epsilon$) adopts the raw price relative and has a different inequality sign. After a certain transformation, PAMR may be written in a similar form, as shown in Table 11.1. However, the prediction functions are different (i.e., OLMAR adopts multiperiod mean reversion, while PAMR exploits single-period mean reversion).

11.3   Algorithms

The preceding formulation is thus convex and straightforward to solve via convex optimization (Boyd and Vandenberghe 2004). We now derive the OLMAR solution as illustrated in Proposition 11.1.

Proposition 11.1 The solution of OLMAR without considering the nonnegativity constraint is

$$\mathbf{b}_{t+1} = \mathbf{b}_t + \lambda_{t+1}(\tilde{\mathbf{x}}_{t+1} - \bar{x}_{t+1}\mathbf{1}),$$

where $\bar{x}_{t+1} = \frac{1}{m}(\mathbf{1} \cdot \tilde{\mathbf{x}}_{t+1})$ denotes the average predicted price relative, and λ_{t+1} is the Lagrangian multiplier calculated as

$$\lambda_{t+1} = \max\left\{0, \frac{\epsilon - \mathbf{b}_t \cdot \tilde{\mathbf{x}}_{t+1}}{\|\tilde{\mathbf{x}}_{t+1} - \bar{x}_{t+1}\mathbf{1}\|^2}\right\}.$$

Proof   *The proof can be found in Appendix B.4.1.*

Following PAMR and CWMR, the above derivation first ignores the nonnegativity constraint (Helmbold et al. 1998). Thus, it is possible that the resulting portfolio goes out of the portfolio simplex domain. To maintain a proper portfolio, we finally

---

Algorithm 11.1: Online portfolio selection with OLMAR.

Input: $\epsilon > 1$: Reversion threshold; $w \geq 1$: Window size; $\alpha \in (0, 1)$: Decaying factor; $\mathbf{x}_1^n$: Market sequence.

Output: S_n: Cumulative wealth after n periods.

begin

    Initialization: $\mathbf{b}_1 = \frac{1}{m}\mathbf{1}$, $S_0 = 1$, $\tilde{\mathbf{x}}_1 = \mathbf{1}$;

    for $t = 1, \ldots, n$ **do**

        Rebalance the portfolio to $\mathbf{b}_t$

        Receive stock price relatives: $\mathbf{x}_t$

        Calculate daily return and cumulative return: $S_t = S_{t-1} \times (\mathbf{b}_t \cdot \mathbf{x}_t)$

        Predict next price relative vector:

$$\tilde{\mathbf{x}}_{t+1} = \begin{cases} \frac{1}{w}\left(1 + \frac{1}{\mathbf{x}_t} + \cdots + \frac{1}{\odot_{i=0}^{w-2}\mathbf{x}_{t-i}}\right) & \text{MAR-1} \\ \alpha\mathbf{1} + (1 - \alpha)\frac{\tilde{\mathbf{x}}_t}{\mathbf{x}_t} & \text{MAR-2} \end{cases}$$

        Update the portfolio:

$$\mathbf{b}_{t+1} = \text{OLMAR}(\epsilon, \tilde{\mathbf{x}}_{t+1}, \mathbf{b}_t)$$

    end

end

Algorithm 11.2: Online Moving Average Reversion: OLMAR(ϵ, $\tilde{\mathbf{x}}_{t+1}$, $\mathbf{b}_t$).

Input: $\epsilon > 1$: Reversion threshold; $\tilde{\mathbf{x}}_{t+1}$: Predicted price relatives; $\mathbf{b}_t$: Current portfolio.

Output: $\mathbf{b}_{t+1}$: Next portfolio.

begin

Calculate the following variables:

$$\bar{x}_{t+1} = \frac{\mathbf{1}^\top \tilde{\mathbf{x}}_{t+1}}{m}, \ \lambda_{t+1} = \max\left\{0, \frac{\epsilon - \mathbf{b}_t \cdot \tilde{\mathbf{x}}_{t+1}}{\|\tilde{\mathbf{x}}_{t+1} - \bar{x}_{t+1}\mathbf{1}\|^2}\right\}$$

Update the portfolio:

$$\mathbf{b}_{t+1} = \mathbf{b}_t + \lambda_{t+1}(\tilde{\mathbf{x}}_{t+1} - \bar{x}_{t+1}\mathbf{1})$$

Normalize $\mathbf{b}_{t+1}$:

$$\mathbf{b}_{t+1} = \arg\min_{\mathbf{b}\in\Delta_m} \|\mathbf{b} - \mathbf{b}_{t+1}\|^2$$

end

---

project the portfolio to the simplex domain (Duchi et al. 2008), which costs linear time.

To this end, we can design the proposed algorithm based on the proposition. The proposed OLMAR procedure is demonstrated in Algorithm 11.1, and the OLPS procedure utilizing the OLMAR algorithm is illustrated in Algorithm 11.2.

11.4 Analysis

The update of OLMAR is straightforward, that is, $\mathbf{b}_{t+1} = \mathbf{b}_t + \lambda_{t+1}(\tilde{\mathbf{x}}_{t+1} - \bar{x}_{t+1}\mathbf{1})$. This second part of the update formula, $+\lambda_{t+1}(\tilde{\mathbf{x}}_{t+1} - \bar{x}_{t+1}\mathbf{1})$, coincides with the general form (Conrad and Kaul 1998, Eq. (1)) of return-based momentum strategies, except the varying λ_{t+1}. Intuitively, the update divides assets into two groups by prediction average. For assets in the group with higher predictions than average, OLMAR increases their proportions; for other assets, OLMAR decreases their proportions. The transferred proportions are related to the surprise of predictions over their average value and the nonnegative Lagrangian multiplier. This is consistent with the normal portfolio selection procedure, that is, to transfer the wealth to assets with a better prospect to grow.

Clearly, the OLMAR update costs linear time per period with respect to m, and the normalization step can also be implemented in linear time (Duchi et al. 2008). To the best of our knowledge, OLMAR's linear time is no worse than any existing algorithms, which can be inferred from Table 10.3.

11.5 Summary

This chapter proposed a novel online portfolio selection (OLPS) strategy named online moving average reversion (OLMAR), which exploits moving average reversion (MAR) via online learning algorithms. The approach can solve the problems of the state of the art caused by the assumption of single-period mean reversion and achieve satisfying results in real markets. It also runs extremely fast and is suitable for large-scale real applications.

In future, we will further explore the theoretical aspect of mean reversion and analyze the behaviors of mean reversion–based portfolios.

Part IV

Empirical Studies

Chapter 12

Implementations

As we have proposed several online portfolio selection (OLPS) algorithms, we are interested in whether they work in real markets. To examine their empirical efficacy, we conducted an extensive set of empirical studies on a variety of real datasets. In our evaluations, we adopted six real datasets, which were collected from several diverse financial markets. The performance metrics include cumulative wealth (return) and risk-adjusted returns (based on volatility risk and drawdown risk). We also compared the proposed algorithms with various existing algorithms. The results clearly demonstrate that the proposed algorithms sequentially surpass the state-of-the-art techniques in terms of either metric.

This chapter is organized as follows. Section 12.1 describes the experimental platform or the OLPS platform. Section 12.2 details the experimental testbed, including six real datasets. Section 12.3 sets up all the proposed algorithms and illustrates several compared approaches. Section 12.4 introduces the performance metrics used for the empirical studies. Finally, Section 12.5 summarizes this chapter.

12.1 The OLPS Platform

To evaluate the performance of a proposed algorithm, researchers and practitioners usually implemented a back-test system, simulating the strategies using historical market data. We also designed a back-test system, named "OLPS", as follows, and Appendix A describes the details of the OLPS toolbox. It implements a framework for back-testing and various algorithms for online portfolio selection. Based on MATLAB®,* it is compatible with Window, Linux, and Mac OS. Figure 12.1 illustrates the structure of the OLPS toolkit, which consists of three parts. The first part on the upper left preprocesses data, that is, it loads a specified dataset and initializes the trading environments, such as log files, timing variable. The second part on the lower level calls OLPS algorithms and simulates the trading process for strategies based on the data prepared in the first part. The third part in the upper right postprocesses the outputs from the second part, that is, it statistically analyzes the returns and calculates some risk-adjusted returns.

---

*More details are available at http://www.mathworks.com

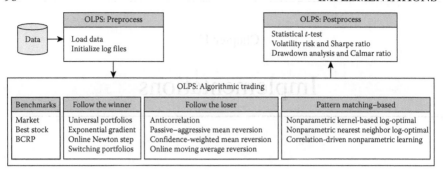

Figure 12.1 *Structure of the OLPS toolbox.*

12.1.1 Preprocess

This step aims to prepare trading environments. As existing datasets are often in MAT files,[*] OLPS accepts datasets in MAT format. The dataset often contains an $n \times m$ matrix, where n denotes the number of trading periods and m refers to the number of assets. It is straightforward to incorporate market feeds[†] from real markets, such that the toolkit can handle real-time data and conduct paper or even real trading.[‡]

12.1.2 Algorithmic Trading

This step conducts simulations based on historical real-market data. In our framework, implementing a new strategy generally requires four files: a start file, a run file, a kernel file, and an expert file. The start (entry) file extracts parameters and call the corresponding run file. The run file simulates a whole trading process and calls its kernel file to construct a portfolio for each period, which is used for rebalancing. The kernel file outputs a final portfolio, while it facilitates the development of meta-algorithms, which effectively combines multiple experts' portfolios. The expert file outputs one portfolio depending on the input data and specific parameters. In case of only one expert, the kernel file is not necessary and directly enters the expert file.

OLPS implements the following OLPS algorithms:

- Benchmarks (Market, Best stock, and BCRP).

- Follow the winner approaches (UP, EG, and ONS): make portfolio decisions following the assumption that the next price relatives (or experts for UP) will follow the previous one.

- Follow the loser approaches (Anticor, PAMR, CWMR, and OLMAR): make portfolio decisions by assuming that next price relatives will revert to previous trends.

---

[*]A full description about MAT files can be found at http://www.mathworks.com/help/pdf_doc/matlab/matfile_format.pdf

[†]For example, Interactive Brokers (http://www.interactivebrokers.com) provides free APIs.

[‡]Both paper and real trading require users to implement an order submission step, while back-test does not.

- Pattern matching–based approaches (B^K, B^{NN}, and CORN): locate a set containing similar price relatives and make optimal portfolios based on the set.
- Others: some are ad hoc algorithms, such as M0/T0.

12.1.3 Postprocess

After the algorithmic trading simulation, this step processes the results by providing the following performance metrics:

- Cumulative return: The most widely used in related studies;
- Volatility and Sharpe ratio: Typically used to measure risk-adjusted return in the investment industry;
- Drawdown and Calmar ratio: Used to measure downside risk and related risk-adjusted return;
- T-test statistics: Tests whether a strategy's return is significantly different from that of the market.

12.2 Data

In our study, we focus on historical daily closing prices in stock markets, which are easy to obtain from public domains (such as Yahoo Finance and Google Finance*), and thus are publicly available to other researchers. Data from other types of markets, such as high-frequency intraday quotes[†] and Forex markets, are either too expensive or hard to obtain and process, and thus may reduce the experimental reproducibility. Summarized in Table 12.1, six real and diverse datasets from several financial markets[‡] are employed.

The first dataset, "NYSE (O)," is one "standard" dataset pioneered by Cover (1991) and followed by others (Helmbold et al. 1998; Borodin et al. 2004; Agarwal et al. 2006; Györfi et al. 2006, 2008). This dataset contains 5651 daily price relatives of 36 stocks[§] in the New York Stock Exchange (NYSE) for a 22-year period from July 3, 1962, to December 31, 1984.

The second dataset is an extended version of the NYSE (O) dataset. For consistency, we collected the latest data in the NYSE from January 1, 1985, to June 30, 2010, a period that consists of 6431 trading days. We denote this new dataset as "NYSE (N)."[¶] Note that the new dataset consists of 23 stocks rather than the previous 36 stocks owing to amalgamations and bankruptcies. All self-collected price

---

*Yahoo Finance: http://finance.yahoo.com; and Google Finance: http://www.google.com/finance

[†]We did evaluate certain algorithms using high-frequency data and weekly data, as in Li et al. (2013).

[‡]All related codes and datasets, including their compositions, are available at http://stevenhoi.org/olps Borodin et al. (2004)'s datasets (NYSE (O), TSE, SP500, and DJIA) are also available at http://www.cs.technion.ac.il/~rani/portfolios/

[§]According to Helmbold et al. (1998), the dataset was originally collected by Hal Stern. The stocks are mainly large cap stocks in NYSE; however, we do no know the criteria of choosing these stocks.

[¶]The dataset before 2007 was collected by Gábor Gelencsér (http://www.cs.bme.hu/~oti/portfolio); we collected the remaining data from 2007 to 2010 via Yahoo Finance.

Table 12.1 *Summary of the six datasets from real markets*

Dataset	Market	Region	Time Frame	# Periods	# Assets
NYSE (O)	Stock	USA	July 3, 1962– December 31, 1984	5651	36
NYSE (N)	Stock	USA	January 1, 1985– June 30, 2010	6431	23
TSE	Stock	CA	January 4, 1994– December 31, 1998	1259	88
SP500	Stock	USA	January 2, 1998– January 31, 2003	1276	25
MSCI	Index	Global	April 1, 2006– March 31, 2010	1043	24
DJIA	Stock	USA	January 14, 2001– January 14, 2003	507	30

relatives are adjusted for splits and dividends, which is consistent with the previous "NYSE (O)" dataset.

The third dataset, "TSE," is collected by Borodin et al. (2004), and it consists of 88 stocks from the Toronto Stock Exchange (TSE) containing price relatives of 1259 trading days, ranging from January 4, 1994, to December 31, 1998. The fourth dataset, SP500, is collected by Borodin et al. (2004), and it consists of 25 stocks with the largest market capitalizations in the 500 SP500 components. It ranges from January 2, 1998, to January 31, 2003, containing 1276 trading days.

The fifth dataset is "MSCI," which is a collection of global equity indices that constitute the MSCI World Index.* It contains 24 indices that represent the equity markets of 24 countries around the world, and it consists of a total of 1043 trading days, ranging from April 1, 2006, to March 31, 2010. The final dataset is the DJIA dataset (Borodin et al. 2004), which consists of 30 Dow Jones composite stocks. DJIA contains 507 trading days, ranging from January 14, 2001, to January 14, 2003.

Besides the six real-market data, in the main experiments (i.e., Experiment 1 in Section 13.1), we also evaluate each dataset in their reversed form (Borodin et al. 2004). For each dataset, we create a reversed dataset, which reverses the original order and inverts the price relatives. We denote these reverse datasets using a '−1' superscript on the original dataset names. In nature, these reverse datasets are quite different from the original datasets, and we are interested in the behaviors of the proposed algorithms on such artificial datasets.

Unlike previous studies, the above testbed covers much longer trading periods from 1962 to 2010 and much more diversified markets, which enables us to examine the behaviors of the proposed strategies under different events and crises. For example, it covers several well-known events in the stock markets, such as the

---

*The constituents of the MSCI World Index are available on MSCI Barra (http://www.mscibarra.com), accessed on 28 May 2010.

dot-com bubble from 1995 to 2000 and the subprime mortgage crisis from 2007 to 2009. The five stock datasets are mainly chosen to test the capability of the proposed algorithms on regional stock markets, while the index dataset aims to test their capability on global indices, which may be potentially applicable to a fund of funds (FOF).* As a remark, although we numerically test the proposed algorithms on stock and exchange traded funds (ETF) markets, we note that the proposed strategies could be generally applied to any type of financial market.

12.3 Setups

In our experiments, we implemented all the proposed approaches: CORN-U, CORN-K, PAMR, PAMR-1, PAMR-2, CWMR-Var, CWMR-Stdev, OLMAR-1, and OLMAR-2. For CWMR algorithms, we only present the results achieved by the deterministic versions. The results of the stochastic versions are presented in Li et al. (2013). Besides individual algorithms, we also designed their buy and hold (BAH) versions whose results can be found on their respective studies (Li et al. 2011b, 2012, 2013; Li and Hoi 2012). Without ambiguity, when referring to CORN, PAMR, CWMR, and OLMAR, we often focus on their representative versions, that is, CORN-U, PAMR, CWMR-Stdev, and OLMAR-1, respectively.

As the proposed algorithms are all online, we follow the existing work and simply set the parameters empirically without tuning for each dataset separately. Note that the best values for these parameters are often dataset dependent, and our choices are not always the best, as we will further evaluate in Section 13.3. Below, we introduce the parameter settings of the proposed algorithms.

For the proposed CORN experts, two possible parameters can affect their performance, that is, the correlation coefficient threshold ρ and the window size w. In our evaluations, we simply fix $\rho = 0.1$ and $W = 5$ for the CORN-U algorithm, which is not always the best. And for the CORN-K algorithm, we first fix $W = 5$, $P = 10$, and $K = 50$, which means choose all experts in the experiments and denote it as "CORN-K1." We also provide "CORN-K2," whose parameters are fixed as $W = 5$, $P = 10$, and $K = 5$.

There are two key parameters in the proposed PAMR algorithms. One is the sensitivity parameter ϵ, and the other is the aggressiveness parameter C. Specifically, for all datasets and experiments, we set the sensitivity parameter ϵ to 0.5 in the three algorithms, and set the aggressiveness parameter C to 500 in both PAMR-1 and PAMR-2, with which the cumulative wealth achieved tends to be stable on most datasets. Our experiments on the parameter sensitivity show that the proposed PAMR algorithms are quite robust with respect to different parameter settings.

CWMR has two key parameters, that is, the confidence parameter ϕ and the sensitivity parameter ϵ. We set the sensitivity parameter ϵ to 0.5 and set the confidence parameter ϕ to 2.0, or equivalently 95% confidence level, in both CWMR-Var and CWMR-Stdev. As the results show, the proposed CWMR algorithm is generally

---

*Note that not every index is tradable through ETFs.

robust with respect to different parameter settings and our choices are not always the best.

For OLMAR, in all cases, we empirically set the mean reversion parameter, that is, $\epsilon = 10$, which provides consistent results. Individually, we set $w = 5$ for OLMAR-1 and $\alpha = 0.5$ for OLMAR-2. As the results show, it is easy to choose satisfying parameters for the proposed OLMAR algorithms.

12.3.1 Comparison Approaches and Their Setups

We compare the proposed algorithms with a number of benchmarks and representative strategies. Below we summarize a list of compared algorithms, all of which provide extensive empirical evaluations in their respective studies. Focusing on empirical studies, we ignore certain algorithms that focus on theoretical analysis and lack thorough empirical evaluations.[*] All parameters are set following their original studies.[†]

1. Market: Market strategy, that is, uniform BAH strategy
2. Best-Stock: Best stock in the market, which is a strategy in hindsight
3. BCRP: Best constant rebalanced portfolios strategy in hindsight
4. UP: Cover's universal portfolios implemented according to Kalai and Vempala (2002), where the parameters are set as $\delta_0 = 0.004$, $\delta = 0.005$, $m = 100$, and $S = 500$
5. EG: Exponential gradient algorithm with the best learning rate $\eta = 0.05$ as suggested by Helmbold et al. (1998)
6. ONS: Online Newton step with the parameters suggested by Agarwal et al. (2006), that is, $\eta = 0, \beta = 1, \gamma = \frac{1}{8}$
7. Anticor: $BAH_{30}(Anticor(Anticor))$ as a variant of Anticor to smooth the performance, which achieves the best performance among the three solutions proposed by Borodin et al. (2004)
8. B^K: Nonparametric kernel-based moving window strategy with $W = 5$, $L = 10$, and threshold $c = 1.0$, which has the best empirical performance according to Györfi et al. (2006)
9. B^{NN}: Nonparametric nearest-neighbor-based strategy with parameters $W = 5$, $L = 10$, and $p_\ell = 0.02 + 0.5\frac{\ell-1}{L-1}$, as the authors suggested (Györfi et al. 2008)

12.4 Performance Metrics

We adopt the most common metric, *cumulative wealth*, to primarily compare different trading strategies. In addition to the cumulative wealth, we also adopt the *annualized Sharpe ratio* (SR) to compare the performance of different trading algorithms. In general, higher values of the cumulative wealth and annualized SR indicate better algorithms. Besides, we also adopt *maximum drawdown* (MDD) and the *Calmar ratio* (CR) for analyzing a strategy's downside risk. The lower the MDD values,

---

[*]Our OLPS platform provides these algorithms.

[†]We can tune their parameters for better performance, but it is beyond the scope of this book.

Table 12.2 *Summary of the performance metrics used in the evaluations*

Criteria	Performance Metrics	
Absolute return	Cumulative wealth (S_n)	Annualized percentage yield
Risk	Annualized standard deviation	Maximum drawdown
Risk-adjusted return	Annualized Sharpe ratio (SR)	Calmar ratio (CR)

the less the strategy's (downside) risk. The higher the CR values, the better the strategy's (downside) risk-adjusted return. We summarize them in Table 12.2 and present their details as follows.

12.5 Summary

A strategy has to be back-tested using historical market data, such that we have confidence that it will continue to be effective in the unseen future markets. This chapter introduces some implementation issues for the empirical studies, including the platform, data, and various setups. In future, we can further extend the online portfolio selection (OLPS) system using real-market feeds and execute the orders using a paper trading account or real trading account. The next chapter will demonstrate the empirical results obtained from the implementation and corresponding back-tests.

Chapter 13

Empirical Results

This chapter introduces the empirical results of the algorithms using the historical market data. These results will demonstrate the effectiveness of these strategies and provide confidence on their practicability in real trading. We also relax some constraints to evaluate their capability in real trading scenarios.

This chapter is organized as follows. Section 13.1 conducts the experiments to evaluate the cumulative wealth for all the algorithms. Section 13.2 shows the experimental results of risk-adjusted returns. Section 13.3 measures the sensitivity of parameters for these algorithms. Section 13.4 relaxes transaction costs and margin buying constraints. Section 13.5 compares the computational times for different algorithms. Section 13.6 further analyzes the behaviors of the proposed algorithms. Finally, Section 13.7 summarizes this chapter and proposes some future directions.

13.1 Experiment 1: Evaluation of Cumulative Wealth

First, we compare the performance of the competing approaches based on their cumulative return, which is the main metric of this study. From the experimental results shown in Table 13.1, we can draw several observations.

First of all, we observe that most online portfolio selection (OLPS) strategies generally perform better than the market and the best stock in a market, which indicates that it is promising to investigate learning algorithms for portfolio selection. Second, although the follow the winner approaches (UP, EG, and ONS) achieve higher cumulative wealth than the market strategy, their performance is significantly less than that of the follow the loser approach (Anticor) or the pattern matching–based strategies (B^K and B^{NN}). Thus, to achieve better investment return, it is more powerful and promising to exploit the latter two approaches. Third, on all original datasets (except the DJIA dataset), the proposed strategies significantly outperform most competitors, including Anticor, B^K, and B^{NN}, which are the state of the art. In particular, the proposed algorithms sequentially beat existing strategies. For example, on the benchmark dataset NYSE (O), the state-of-the-art performance is 3.35E+11 achieved by B^{NN}. Our proposed algorithms achieve much better performances of 1.48E+13, 5.14E+15, 6.51E+15, and 3.68E+16 for CORN, PAMR, CWMR, and OLMAR, respectively.

Table 13.1 *Cumulative wealth achieved by various trading strategies on the six datasets and their reversed datasets*

Algorithms	NYSE (O)	NYSE (N)	TSE	SP500	MSCI	DJIA
Market	14.50	18.06	1.61	1.34	0.91	0.76
Best-stock	54.14	83.51	6.28	3.78	1.50	1.19
BCRP	250.60	119.81	6.78	4.07	1.51	1.24
UP	26.68	31.49	1.60	1.62	0.92	0.81
EG	27.09	31.00	1.59	1.63	0.93	0.81
ONS	109.19	21.59	1.62	3.34	0.86	1.53
Anticor	2.41E+08	6.21E+06	39.36	5.89	3.22	**2.29**
B^K	1.08E+09	4.64E+03	1.62	2.24	2.64	0.68
B^{NN}	3.35E+11	6.80E+04	2.27	3.07	13.47	0.88
CORN-U	1.48E+13	5.37E+05	3.56	6.35	26.10	0.84
CORN-K1	3.19E+13	1.94E+05	1.65	4.64	16.32	0.79
CORN-K2	6.10E+13	4.86E+05	1.74	9.12	**80.41**	0.82
PAMR	5.14E+15	1.25E+06	264.86	5.09	15.23	0.68
PAMR-1	5.13E+15	1.26E+06	260.26	5.08	15.51	0.69
PAMR-2	4.88E+15	1.36E+06	249.95	5.00	16.87	0.71
CWMR-Var	6.51E+15	1.44E+06	328.61	5.94	17.27	0.69
CWMR-Stdev	6.49E+15	1.41E+06	332.62	5.90	17.28	0.68
OLMAR-1	3.68E+16	2.54E+08	424.80	5.83	16.39	2.12
OLMAR-2	**1.02E+18**	**4.69E+08**	**732.44**	**9.59**	22.51	1.16

Algorithms	NYSE (O)$^{-1}$	NYSE (N)$^{-1}$	TSE^{-1}	SP500^{-1}	MSCI^{-1}	DJIA^{-1}
Market	0.12	1.27	1.67	0.88	1.26	1.44
Best-stock	0.33	24.59	37.65	1.65	3.45	2.77
BCRP	2.86	56.60	58.61	1.91	3.45	2.98
UP	0.23	0.3	1.18	1.10	1.26	1.54
EG	0.22	0.38	1.21	1.08	1.27	1.53
ONS	0.84	1.01	1.62	2.97	1.73	2.35
Anticor	1.38E+03	4.26E+04	7.24	9.64	6.31	4.58
B^K	2.77E+07	162.74	8.81	1.01	4.47	1.43
B^{NN}	4.60E+09	3.57E+04	**66.09**	1.89	30.06	1.85
CORN-U	1.74E+10	8.01E+03	53.06	1.81	36.05	1.83
CORN-K1	4.99E+09	6.79E+03	12.88	1.67	17.87	1.75
CORN-K2	**3.19E+10**	7.27E+03	40.87	2.63	**66.64**	1.66
PAMR	2.03E+04	3.07E+04	2.67	7.42	40.33	6.61
PAMR-1	2.02E+04	3.09E+04	2.68	7.43	39.82	6.62
PAMR-2	2.11E+04	3.21E+04	2.75	7.32	39.83	6.65
CWMR-Var	1.67E+04	6.35E+04	4.04	8.09	40.46	6.90
CWMR-Stdev	1.66E+04	6.49E+04	4.05	8.07	40.42	6.91
OLMAR-1	2.07E+04	**3.99E+07**	2.90	18.40	42.25	**9.56**
OLMAR-2	3.41E+04	2.26E+07	7.04	**40.99**	51.51	8.80

Note: Numbers in **bold** indicate the best results on the corresponding datasets.

This observation supports the effectiveness of proposed algorithms, and, to the best of our knowledge, no one has ever claimed such a fantastic performance.

Fourthly, it is promising to see that all pattern matching–based algorithms, especially CORN, have better performance than the benchmarks on most datasets. And the proposed CORN significantly outperforms the existing pattern matching–based algorithms, including B^K and B^{NN}, validating its motivating idea of improving the matching process. Fifthly, the encouraging results achieved by the last three strategies (PAMR, CWMR, and OLMAR) validate the importance of exploiting mean reversion in financial markets via an effective learning algorithm.

In addition, we can see that most algorithms perform poorly on the DJIA dataset, including CORN, PAMR, and CWMR. While the failure of CORN is still unexplainable, the failure of the two mean reversion algorithms indicates that the motivating (single-period) mean reversion may not exist in the dataset, as analyzed in Sections 10.1 and 11.1.2. While OLMAR is proposed to explore (multiple-period) moving average reversion, it can achieve much better performance, which thus validates its motivating idea.

On the reversed datasets, though not as shiny as the original datasets, the proposed algorithms also perform excellently. In all cases, the proposed algorithms not only beat the benchmarks, including the market and BCRP, but also achieve the best performance. Note that these reversed datasets are artificial datasets, which never exist in real markets. However, algorithms' behaviors on these datasets still provide strong evidence that the proposed algorithms can effectively exploit the markets and outperform the benchmarks and the state of the art.

Besides the final cumulative wealth, we are also interested in examining how the cumulative wealth changes over the entire trading periods. Figure 13.1 shows the trends of cumulative wealth by the proposed algorithms and four existing algorithms (two benchmarks and two state-of-the-art algorithms). Note that PAMR and CWMR almost overlap on the figures; thus, we only present the trends of PAMR. Clearly, the proposed strategies consistently surpass the benchmarks and the competing strategies over the entire trading periods on most datasets (except DJIA dataset), which again validates the efficacy of the proposed techniques.

Finally, to measure whether such excess returns can be obtained by simple luck, we conduct a statistical t-test as described in Section 12.4. Table 13.2 shows the statistical results on the four proposed algorithms. The results clearly show that the observed excess return is impossible to obtain by simple luck on most datasets. To be specific, on datasets except DJIA, the probabilities for achieving the excess return by luck are almost 0. On the DJIA dataset, though PAMR and CWMR have a probability of 40% to achieve the excess return by luck, OLMAR has a probability of only 1.69%. Nevertheless, the results show that the proposed strategies are promising and reliable to achieve high returns with high confidence.

13.2   Experiment 2: Evaluation of Risk and Risk-Adjusted Return

We now evaluate the volatility risk and drawdown risk, and the risk-adjusted return in terms of an annualized Sharpe ratio (SR) and Calmar ratio (CR). Figure 13.2 shows

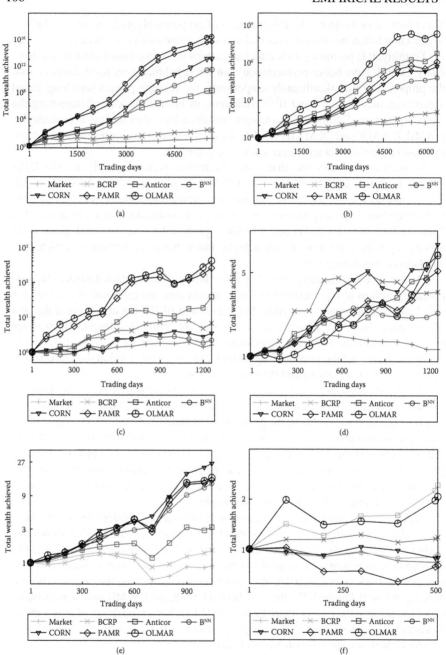

Figure 13.1 *Trends of cumulative wealth achieved by various strategies on the six datasets: (a) NYSE (O); (b) NYSE (N); (c) TSE; (d) SP500; (e) MSCI; and (f) DJIA.*

Table 13.2 *Statistical t-test of the proposed algorithms on the six datasets*

Algorithms	Statistics	NYSE (O)	NYSE (N)	TSE	SP500	MSCI	DJIA
Market	Size	5651	6431	1259	1276	1043	507
	MER	0.0005	0.0005	0.0004	0.0003	0.0000	−0.0004
CORN	WR	53.78%	52.64%	51.55%	52.82%	62.22%	51.08%
	MER	0.0058	0.0023	0.0014	0.0018	0.0033	−0.0002
	α	0.0052	0.0017	0.0008	0.0014	0.0032	0.0002
	β	1.2351	1.0552	1.5050	1.3096	0.8622	0.9047
	t-statistics	13.8069	8.2069	1.2119	2.9073	9.8834	0.3466
	p-value	0.0000	0.0000	0.1129	0.0019	0.0000	0.3645
PAMR	WR	55.87%	51.75%	56.87%	53.37%	59.25%	51.87%
	MER	0.0069	0.0026	0.0054	0.0017	0.0029	−0.0003
	α	0.0063	0.0021	0.0049	0.0013	0.0029	0.0002
	β	1.2095	1.1241	1.4982	1.2375	1.1177	1.2393
	t-statistics	15.7829	5.9979	3.9241	2.0020	6.1358	0.2195
	p-value	0.0000	0.0000	0.0000	0.0227	0.0000	0.4132
CWMR	WR	56.17%	52.08%	56.00%	53.92%	59.44%	51.08%
	MER	0.0070	0.0027	0.0057	0.0019	0.0030	−0.0003
	α	0.0064	0.0021	0.0051	0.0015	0.0030	0.0002
	β	1.2139	1.1325	1.5139	1.2512	1.1161	1.2476
	t-statistics	15.9510	5.9496	3.9190	2.1806	6.4078	0.2482
	p-value	0.0000	0.0000	0.0000	0.0147	0.0000	0.4020
OLMAR	WR	56.91%	53.13%	55.12%	51.49%	58.39%	52.47%
	MER	0.0074	0.0036	0.0061	0.0019	0.0030	0.0020
	α	0.0068	0.0030	0.0056	0.0015	0.0030	0.0025
	β	1.2965	1.1768	1.5320	1.2854	1.1763	1.2627
	t-statistics	15.2405	7.3704	3.4583	1.9423	1.1763	1.2627
	p-value	0.0000	0.0000	0.0003	0.0262	0.0000	0.0169

Note: MER denotes mean excess return, which equals the mean of daily returns over a risk-free return. WR denotes winning ratio, which is the ratio of trading periods with a higher return than the market.

the evaluation results on the six datasets. In addition to the proposed four algorithms, we also plot two benchmarks (Market and BCRP) and two state-of-the-art algorithms (Anticor and B^{NN}). In particular, Figure 13.2a and 13.2b depicts the volatility risk (standard deviation of daily returns) and the drawdown risk (maximum drawdown) on the six stock datasets. Figure 13.2c and 13.2d compares their corresponding SRs and CRs.

In the preceding results on cumulative wealth, we find that the proposed methods achieve the highest cumulative return on most original datasets. However, high return is associated with high risk, as no real financial instruments can guarantee high return

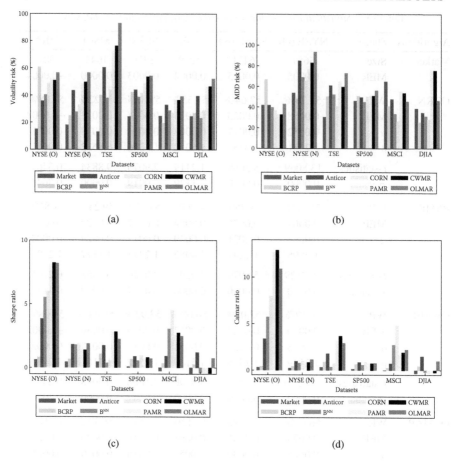

Figure 13.2 *Risk and risk-adjusted performance of various strategies on the six datasets. In each diagram, the rightmost four bars represent the results of our proposed strategies: (a) volatility risk; (b) drawdown risk; (c) Sharpe ratio; and (d) Calmar ratio.*

without high risk.* The volatility risk in Figure 13.2a shows that the proposed four methods almost achieve the highest risk in terms of volatility risk on most datasets. On the other hand, the drawdown risk in Figure 13.2b shows that the proposed methods also achieve high drawdown risk in most datasets. These results validate the notion that high return is often associated with high risk.

To further evaluate the return and risk, we examine the risk-adjusted return in terms of an annualized SR and CR. The results in Figure 13.2c and 13.2d clearly show that CORN, PAMR, and CWMR achieve excellent performance in most cases,

---

*It is true for the long-only portfolio, which is our setting. However, such a statement may be suspect in regard to long-short portfolios.

except the DJIA dataset; and OLMAR achieves excellent performance in all datasets. These encouraging results show that the proposed methods are able to reach a good trade-off between return and risk, even though we do not explicitly consider risk in the method formulations.*

13.3 Experiment 3: Evaluation of Parameter Sensitivity

In the following four subsections, we evaluate how different choices of parameters affect the proposed four strategies.

13.3.1 CORN's Parameter Sensitivity

The proposed CORN has two parameters, that is, the correlation coefficient threshold ρ and the window size for the experts w (or W).

First, let us see the effects of ρ with fixed W, in Figure 13.3. Clearly, the figures validate the preliminary analysis in Section 8.4. In general, CORN achieves the best performance when ρ is around 0, as the figures often peak around 0 or some small positive values; and, when ρ approaches -1 or 1, CORN's performance degrades. Although CORN does not perform well on the TSE and DJIA datasets, on which the cumulative wealth is often less than the BCRP strategy, it significantly outperforms the two benchmarks on other datasets. Based on the above observation, choosing a satisfying ρ for CORN is straightforward, as some small positive values often give good performance on all datasets.

We also examine the effects of W with fixed ρ in Figure 13.4. Note that here CORN denotes the CORN experts with a specified w and CORN-U denotes the uniform combination of CORN experts with w from 1 to W. Although the cumulative wealth achieved by CORN experts fluctuates with different w's, CORN-U's cumulative wealth is much more robust with respect to W. Such an observation validates the effectiveness of the proposed CORN-U and eases the selection of a satisfying W.

13.3.2 PAMR's Parameter Sensitivity

In this section, we examine PAMR's parameters, that is, the mean reversion threshold ϵ for the three algorithms and the aggressiveness parameter C for the two variants.

First, we examine the effect of ϵ on PAMR's cumulative wealth. As ϵ is greater than 1, PAMR degrades to uniform constant rebalanced portfolios (CRP) strategy, and the wealth stabilizes at a constant value achieved by uniform CRP. Thus, we show the effect of ϵ in the range of [0, 1.5]. Figure 13.5 shows the cumulative wealth achieved by PAMR with varying ϵ and two benchmarks, that is, Market and BCRP. Results on most datasets, except the DJIA dataset, show that the cumulative wealth achieved by PAMR consistently grows as ϵ approaches 0. That is, the smaller the threshold, the higher the cumulative wealth is, which validates that the motivating mean reversion does exist on most stock markets. Moreover, in most cases, the cumulative wealth tends to stabilize as ϵ crosses certain dataset-dependent thresholds. As stated before,

---

*We will study it in future.

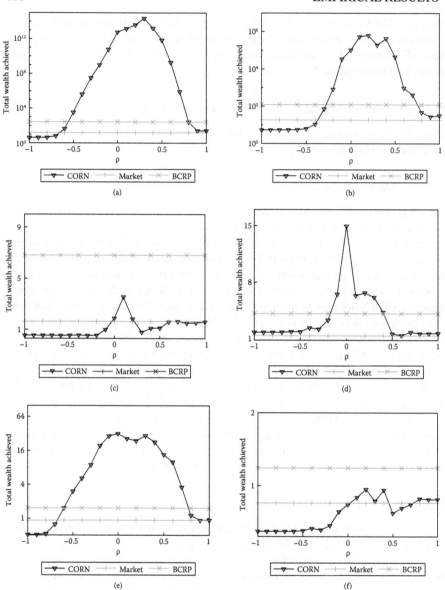

Figure 13.3 *Parameter sensitivity of CORN-U with respect to ρ with fixed W (W = 5): (a) NYSE (O); (b) NYSE (N); (c) TSE; (d) SP500; (e) MSCI; and (f) DJIA.*

we choose $\epsilon = 0.5$ in the experiments, with which the cumulative wealth stabilizes in most cases. Contrarily, on the DJIA dataset, as ϵ approaches 0, the cumulative wealth achieved by PAMR drops. Such phenomena can be interpreted to mean that the motivating (single-period) mean reversion does not exist on the dataset, at least in

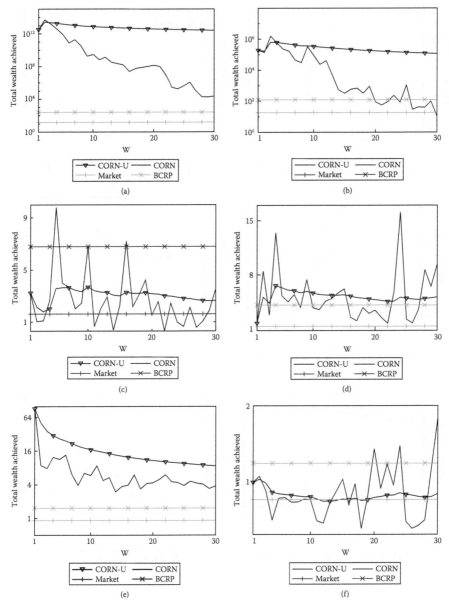

Figure 13.4 *Parameter sensitivity of CORN-U with respect to w (W) with fixed ρ (ρ = 0.1): (a) NYSE (O); (b) NYSE (N); (c) TSE; (d) SP500; (e) MSCI; and (f) DJIA.*

the sense of our motivation. We also note that, on some datasets, PAMR with $\epsilon = 0$ achieves the best. Though $\epsilon = 0$ means moving more weights to underperforming stocks, it may not mean moving everything to the worst stock. On the one hand, the objectives in the formulations would prevent the next portfolio from being far from the

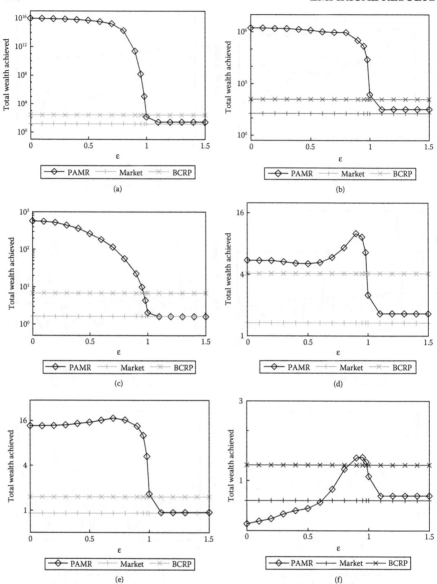

Figure 13.5 *Parameter sensitivity of PAMR with respect to ε: (a) NYSE (O); (b) NYSE (N); (c) TSE; (d) SP500; (e) MSCI; and (f) DJIA.*

last portfolio. On the other hand, PAMR-1 and PAMR-2 are designed to alleviate the huge changes. In summary, the experimental results indicate that the proposed PAMR is robust with respect to the mean reversion sensitivity parameter, in most cases.

Second, we evaluate the other important parameter for both PAMR-1 and PAMR-2, that is, the aggressiveness parameter C. Figure 13.6 shows the effects on the

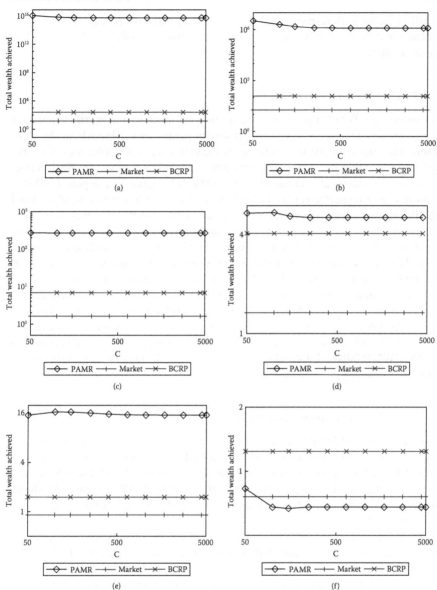

Figure 13.6 *Parameter sensitivity of PAMR-1 (or PAMR-2) with respect to C with fixed* ε *(*ε *= 0.5): (a) NYSE (O); (b) NYSE (N); (c) TSE; (d) SP500; (e) MSCI; and (f) DJIA.*

cumulative wealth by varying C from 50 to 5000 on PAMR-1, with fixed ε = 0.5. We only show the figures on PAMR-1, as the effects on PAMR-2 are similar to those on PAMR-1. Clearly, the proposed PAMR is insensitive to C, and a wide range of values correspond to the highest cumulative wealth. This again exhibits that the proposed

PAMR is robust with respect to its parameters. Similarly, the figure on DJIA again indicates that the mean reversion effect may not exist on the dataset, in the sense of our motivation.

13.3.3 CWMR's Parameter Sensitivity

The proposed CWMR algorithms contain two parameters, that is, the confidence parameter ϕ and the mean reversion sensitivity parameter ϵ. Throughout the algorithms, the mean reversion sensitivity parameter decisively influences the final performance, while the confidence parameter has little effect on the final performance, whose effects are similar to C on PAMR-1/2. Figure 13.7 depicts CWMR's robustness with respect to ϵ, plus the final cumulative wealth achieved by Market and BCRP. The results first show that the final cumulative wealth increases as the sensitivity parameter decreases and stabilizes after ϵ falls below certain data-dependent thresholds, which means that the mean reversion idea has been completely exploited. Then, the results again verify that the mean reversion trading idea works in the financial markets and the proposed CWMR algorithm can successfully exploit it, which generates significant final cumulative wealth on most datasets. Moreover, as analyzed in Section 10.4, CWMR degrades to uniform CRP strategy when ϵ is larger than 1. Needless to say, our empirical setting $\epsilon = 0.5$ is not the best one; however, the proposed CWMR still significantly surpasses existing approaches. Finally, similar to PAMR, CWMR fails on DJIA, which still indicates that the motivating (single-period) mean reversion may not exist on the dataset.

13.3.4 OLMAR's Parameter Sensitivity

Now we evaluate OLMAR's sensitivity to its parameters, that is, ϵ for both OLMAR-1 and OLMAR-2, and w and α for OLMAR-1 and OLMAR-2, respectively. Figure 13.8 shows OLMAR-1's sensitivity to ϵ with fixed $w = 5$. Since OLMAR-2 and OLMAR-1 have the similar figures for ϵ, we only list its effects on OLMAR-1. Figure 13.9 shows its sensitivity to w with fixed $\epsilon = 10$, and Figure 13.10 shows OLMAR-2's sensitivity to α with fixed $\epsilon = 10$.

From Figure 13.8, we can observe that, in general, the cumulative wealth sharply increases if ϵ approaches 1 and flattens if ϵ crosses a threshold. From Figure 13.9, we can see that as w increases, the performance initially increases, spikes at a data-dependent value, and then decreases. Regardless, its performance with most choices of ϵ and w is much better than that of the market and BCRP. To smooth the volatility of its performance, Figure 13.9 also shows OLMAR's BAH versions (Li and Hoi 2012) by combining a set of OLMAR experts with varying w's. We can see that the BAH version provides a much smoother cumulative return than its underlying experts. Note that on DJIA, OLMAR performs much better than PAMR/CWMR as ϵ varies, which validates the motivating multiperiod mean reversion. From Figure 13.10, we can find that on most datasets, α provides significant high cumulative wealth in a wide range of values, except the two extreme endpoints, that is, 0 and 1. If $\alpha = 1$, then all expected price relatives are always 1 and OLMAR-2 outputs $\mathbf{b}_{t+1} = \mathbf{b}_t$, which is

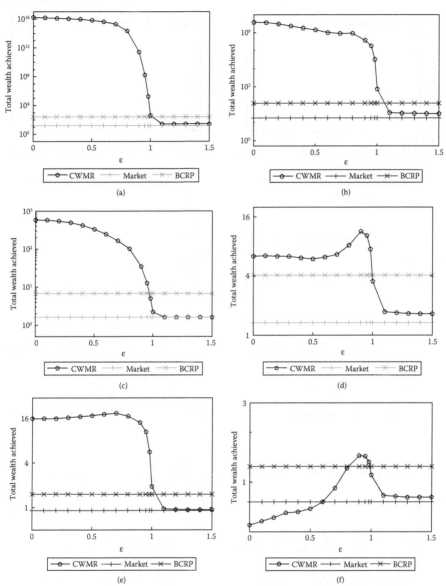

Figure 13.7 *Parameter sensitivity of CWMR with respect to ϵ: (a) NYSE (O); (b) NYSE (N); (c) TSE; (d) SP500; (e) MSCI; and (f) DJIA.*

initialized to uniform portfolio. If $\alpha = 0$, then its expected price relative vector equals $\tilde{\mathbf{x}}_{t+1} = \frac{1}{\bigodot_{i=1}^{t} \mathbf{x}_i}$. Such price relatives inversely relate to all of an asset's historical price relatives and produce bad results.

Nevertheless, all the above observations show that OLMARs' performance is robust to their parameters, and it is convenient to choose satisfying parameters.

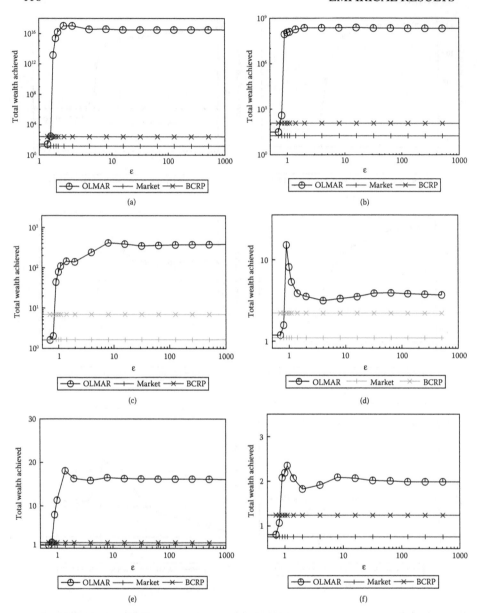

Figure 13.8 *Parameter sensitivity of OLMAR-1 with respect to ϵ with fixed w ($w = 5$): (a) NYSE (O); (b) NYSE (N); (c) TSE; (d) SP500; (e) MSCI; and (f) DJIA.*

13.4 Experiment 4: Evaluation of Practical Issues

For real-world portfolio management, there are some important practical issues, including transaction costs and margin buying. In this section, we examine how the two issues affect the proposed strategies.

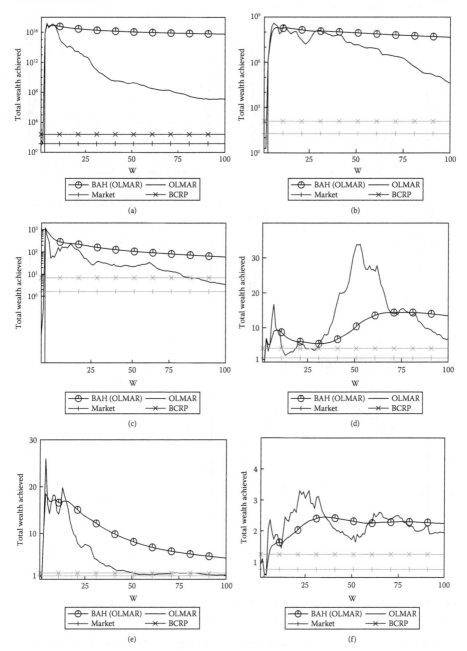

Figure 13.9 *Parameter sensitivity of OLMAR-1 with respect to w with fixed ε (ε = 10): (a) NYSE (O); (b) NYSE (N); (c) TSE; (d) SP500; (e) MSCI; and (f) DJIA.*

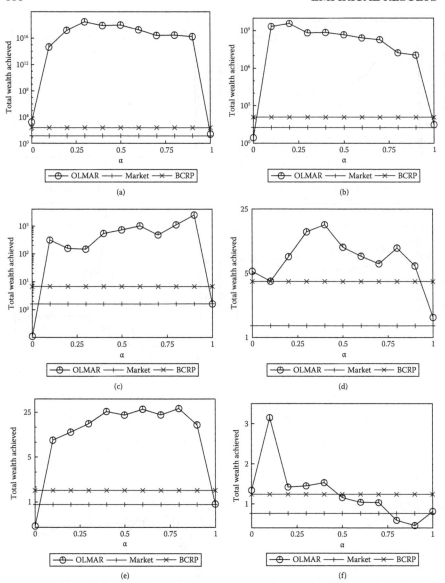

Figure 13.10 *Parameter sensitivity of OLMAR-2 with respect to* α *with fixed* ε *(ε = 10): (a) NYSE (O); (b) NYSE (N); (c) TSE; (d) SP500; (e) MSCI; and (f) DJIA.*

First, the transaction cost is an important and unavoidable issue that should be addressed in practice. To test the effects of transaction cost on the proposed strategies, we adopt the *proportional transaction cost* model stated in Section 2.2. Figure 13.11 depicts the effects of proportional transaction cost when the algorithms are applied on the six datasets, where the transaction cost rate γ varies from 0% to 1%. We present only the results achieved by three representative algorithms (CORN, PAMR,

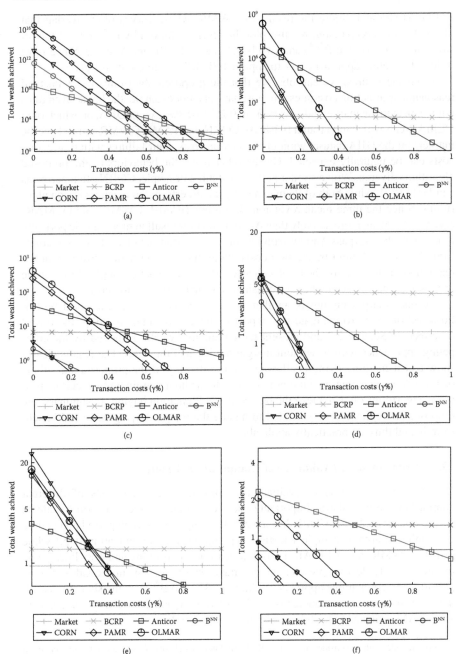

Figure 13.11 *Scalability of the proposed strategies with respect to the transaction cost rate (γ): (a) NYSE (O); (b) NYSE (N); (c) TSE; (d) SP500; (e) MSCI; and (f) DJIA.*

and OLMAR) and ignore the results of CWMR, whose curves often overlap that of PAMR. For comparison, we also plot the results achieved by two state-of-the-art strategies (Anticor and B^{NN}) and two benchmarks (BCRP and Market). Since most follow the winner approaches try to approach BCRP, we ignore their figures.

From the figures, we can observe that the proposed algorithms can withstand reasonable transaction cost rates, on most datasets. For example, the break-even rates with respect to the market index vary from 0.2% to 0.8%, except DJIA, on which only OLMAR can withstand around 0.3%. As CORN and PAMR/CWMR fail to beat the markets on the DJIA dataset without transaction costs, their failures with transaction costs can be naturally expected. On the other hand, the behaviors of the proposed algorithms diverge. With a similar pattern-matching principle, CORN often performs similar to B^{NN}, while both of them generally underperform the mean reversion algorithms. Since the three mean reversion algorithms (PAMR, CWMR, and OLMAR) revert to the mean more actively than Anticor and thus result in more drastic portfolio rebalances, they surpass Anticor with low or medium transaction costs and underperform Anticor with high transaction costs. Note that the transaction cost rate in the real market is low*; thus, the results clearly indicate the practical applicability of the proposed strategies even when we consider reasonable transaction costs.

Second, margin buying is another practical concern for a real-world portfolio selection task. To evaluate the impact of margin buying, we adopt the model described in Section 2.2 and present the cumulative wealth achieved by the competing approaches with or without margin buying in Table 13.3. The results clearly show that if margin buying is allowed, the profitability of the proposed algorithms on most datasets increases. Similar to the results without margin buying, certain proposed algorithms often achieve the best results with margin buying. In summary, the proposed strategies can be extended to handle the margin-buying issue and benefit from it, and thus are practically applicable.

13.5 Experiment 5: Evaluation of Computational Time

Our next experiment is to evaluate the computational time costs of different approaches, which is also an important issue in developing a practical trading strategy. As previously analyzed, CORN has a batch-learning step on each period and is time consuming in both its sample selection step and portfolio optimization step,[†] while PAMR, CWMR, and OLMAR are online learning algorithms and cost linear time per iteration. Table 13.4 presents the computational time cost (in seconds) of three performance-comparable approaches (Anticor, B^K, and B^{NN}) on the six datasets. All the experiments were conducted on an Intel Core 2 Quad 2.66 GHz processor with 4 GB RAM, using MATLAB® 2009b on Windows XP.[‡]

---

*For example, without considering taxes and bid–ask, Interactive Broker (www.interactivebrokers.com) charges $0.005 per share. Since the average price of Dow Jones Composites is around $50.00 (as of June 2011), the transaction cost rate is about 0.01%.

[†]In its MATLAB implementation, the latter step costs more than 80% of the total time.

[‡]We use MATLAB function tic/toc to measure the time. There are preprocessing (such as data loader, variable initialization, etc.) and postprocessing (such as result analysis, etc.), whose time is all excluded from the time statistics in Table 13.4.

Table 13.3 *Cumulative wealth achieved by various strategies on the six datasets without and with margin loans (MLs)*

Algorithms	NYSE (O)		NYSE (N)		TSE	
	No ML	With ML	No ML	With ML	No ML	With ML
Market	14.50	15.75	18.06	17.68	1.61	1.71
Best-stock	54.14	54.14	83.51	173.18	6.28	10.53
BCRP	250.6	3755.09	120.32	893.63	6.78	21.23
UP	27.41	62.99	31.49	57.03	1.60	1.69
EG	27.09	63.28	31.00	55.55	1.59	1.68
ONS	109.19	517.21	21.59	228.37	1.62	0.88
Anticor	2.41E+08	1.05E+15	6.21E+06	5.41E+09	39.36	18.69
B^K	1.08E+09	6.29E+15	4.64E+03	3.72E+06	1.62	1.53
B^{NN}	3.35E+11	3.17E+20	6.80E+04	5.58E+07	2.27	2.17
CORN	1.48E+13	6.59E+25	5.37E+05	7.31E+07	3.56	5.00
PAMR	5.14+15	5.57E+25	1.25E+06	1.12E+09	264.86	720.42
CWMR	6.49E+15	6.59E+25	1.41E+06	7.31E+07	332.62	172.36
OLMAR	3.68E+16	5.67E+30	2.54E+08	1.73E+12	424.80	31.63

Algorithms	SP500		MSCI		DJIA	
	No ML	With ML	No ML	With ML	No ML	With ML
Market	1.34	1.03	0.91	0.69	0.76	0.59
Best-stock	3.78	3.78	1.50	1.50	1.19	1.19
BCRP	4.07	6.48	1.51	1.54	1.24	1.24
UP	1.62	1.75	0.92	0.71	0.81	0.66
EG	1.63	1.70	0.93	0.72	0.81	0.65
ONS	3.34	7.76	0.86	0.33	1.53	2.21
Anticor	5.89	10.73	3.22	3.40	2.29	2.89
B^K	2.24	1.88	2.64	6.56	0.68	0.56
B^{NN}	3.07	3.29	14.47	150.49	0.88	0.67
CORN	6.35	14.59	26.10	835.08	0.84	0.55
PAMR	5.09	15.91	15.23	68.83	0.68	0.84
CWMR	5.90	23.50	17.28	76.29	0.68	0.88
OLMAR	5.83	5.60	16.39	57.79	2.12	1.46

From the results, we can clearly see that CORN and the state-of-the-art algorithms have high costs, and in all cases the proposed PAMR, CWMR, and OLMAR take significantly less computational time than others. Even though the computational time in daily back-tests, especially per trading day, is small, it is important in certain scenarios such as high-frequency trading (Aldridge 2010), where transactions may occur in fractions of a second. Nevertheless, the results obviously demonstrate the computational efficiency of three proposed mean reversion strategies, which further enhances their real-world large-scale applicability.

Table 13.4 *Computational time cost (in seconds) on the six real datasets*

Algorithms	NYSE (O)	NYSE (N)	TSE	SP500	MSCI	DJIA
Anticor	2.57E+03	1.93E+03	2.15E+03	387	306	175
B^K	7.89E+04	5.78E+04	6.35E+03	1.95E+03	2.60E+03	802
B^{NN}	4.93E+04	3.39E+04	1.32E+03	2.91E+03	2.55E+03	1.28E+03
CORN	8.78E+03	1.03E+04	1.59E+03	563	444	172
PAMR	8	7	2	1.1	1.0	0.3
CWMR	12	11	3	1.4	1.3	0.5
OLMAR	4	3	0.7	0.6	0.5	0.3

13.6   Experiment 6: Descriptive Analysis of Assets and Portfolios

Existing experiments on related studies (refer to Experiments 1 to 5) focus on comparing different algorithms based on various preceding aspects. In this section, we perform a preliminary analysis of the behaviors of asset returns and portfolios, which may reflect some insights into future study. While the analysis on different datasets is similar, we focus on the standard benchmark dataset, NYSE (O) (Cover 1991).* We also append the data statistics and top five average allocations of our strategies and the state-of-the-art algorithms on other datasets,[†] such as Appendix C.

Before analyzing their portfolios, we list some descriptive statistics on NYSE (O), including each asset's cumulative return over the whole periods, their (arithmetic) return mean and standard deviation, and their autocorrelation with lag 1, in Table 13.5.

Then, we plot some representative approaches' mean weights and standard deviations in Figure 13.12, and list the top five average weights in Table 13.6. First, let us analyze the BCRP strategy, whose portfolio has the same weights for every period and thus has zero standard deviation. BCRP is essentially different from the best stock strategy (asset #30), as the weight on the stock is zero. Interestingly, BCRP focuses on the five most volatile stocks (refer to the highest Std values in Table 13.5), which means that the portfolio selections are undiversified and verifies the "volatility pumping" (Luenberger 1998) nature. Even though asset #23 does not perform as good as most assets, its high volatility makes it the second weighted asset. This shows that exploiting volatile stocks, even though some of them may perform poorly, can give good performance.

For both EG and ONS, their portfolios have much lower volatility than other strategies. In particular, EG's portfolios always slightly drift around the initial uniform portfolio (for NYSE (O), $\frac{1}{36}\mathbf{1}$). Such a phenomenon can be explained by its learning rate ($\lambda > 0$), which has to be small such that the algorithm is universal. However, decreasing the learning rate ($\lambda \to 0$) ultimately approaches the algorithm to uniform CRP.[‡] Our observation on EG's portfolios verifies the previous analysis on its parameter, in Section 4.2.

---

*Due to the table constraints, we use indices to represent individual assets, whose symbols are available at http://stevenhoi.org/olps

[†]We ignore TSE, which has too much assets (m = 88) to show.

[‡]Uniform CRP will be constant at uniform portfolio. This is approachable but not achievable as $\lambda > 0$.

Table 13.5 *Some descriptive statistics on the NYSE (O) dataset*

Stat.	1	2	3	4	5	6	7	8	9
Cum	13.10	4.35	16.10	16.90	13.36	**52.02**	8.76	3.07	13.71
Mean	1.0005	1.0004	1.0006	1.0006	1.0006	1.0010	1.0005	1.0003	1.0011
Std	0.0135	0.0171	0.0128	0.0170	0.0140	**0.0257**	0.0156	0.0136	**0.0371**
Ac	0.1344	0.1206	0.0817	0.0952	0.0927	0.0378	0.0615	0.0479	**−0.0217**

Stat.	10	11	12	13	14	15	16	17	18
Cum	14.16	10.70	6.85	7.86	6.75	7.64	**32.65**	**30.61**	12.21
Mean	1.0005	1.0006	1.0005	1.0005	1.0004	1.0004	1.0009	1.0008	1.0005
Std	0.0115	0.0175	0.0159	0.0138	0.0137	0.0130	0.0224	0.0202	0.0134
Ac	0.1114	0.1312	0.0455	0.0766	0.0637	0.0744	0.0449	0.0626	**0.0064**

Stat.	19	20	21	22	23	24	25	26	27
Cum	4.81	8.92	17.22	10.36	4.13	6.21	4.31	22.92	14.43
Mean	1.0004	1.0010	1.0006	1.0005	1.0015	1.0004	1.0005	1.0010	1.0006
Std	0.0146	**0.0346**	0.0151	0.0148	**0.0505**	0.0149	0.0230	**0.0313**	0.0142
Ac	0.1301	0.0243	0.0956	0.1047	**−0.2089**	−0.0042	0.0226	**−0.0915**	0.1002

Stat.	28	29	30	31	32	33	34	35	36
Cum	5.98	15.21	**54.14**	6.98	16.20	**43.13**	4.25	6.54	5.39
Mean	1.0004	1.0006	1.0008	1.0004	1.0006	1.0008	1.0004	1.0005	1.0004
Std	0.0139	0.0161	0.0153	0.0117	0.0159	0.0174	0.0143	0.0178	0.0143
Ac	0.0858	0.0697	0.1004	0.0870	0.0880	0.1024	0.0873	0.0626	0.0257

Note: "Cum" denotes the cumulative return (product of all price relatives) of an asset. "Mean" refers to one asset's arithmetic mean, and "Std" denotes the asset's standard deviation. "Ac" denotes the autocorrelation (with lag 1) of an asset. Numbers in **bold** denote the top five in the corresponding rows.

The three pattern matching–based approaches (B^K, B^{NN}, and CORN) have similar patterns in their allocation weights, while their top five allocations vary. In general, their volatilities are much higher than EG and ONS. Their concentration on asset #23, which has the highest weight, confirms the observation (Györfi et al. 2006;

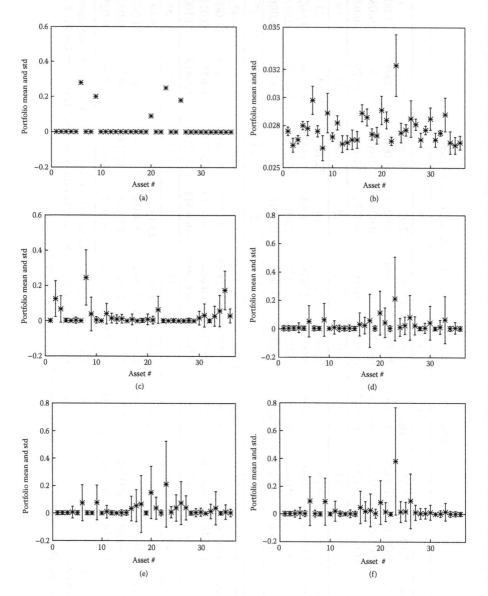

Figure 13.12 *Distributions of portfolio weights. The x-axis denotes indices of assets, and the y-axis is each asset's average weight. For each asset, the center of an error bar denotes its portfolio mean (over 5651 trading days), and vertical lines denote its standard deviations: (a) BCRP; (b) EG; (c) ONS; (d)* B^K; *(e)* B^{NN}; *and (f) CORN.* (Continued)

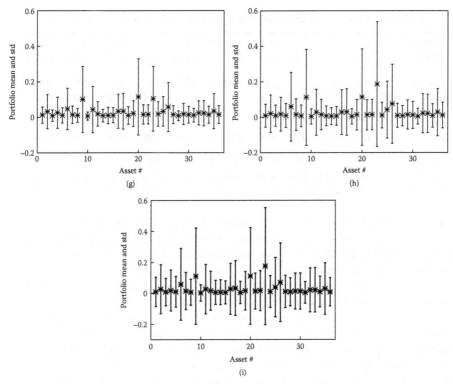

Figure 13.12 (Continued) *Distributions of portfolio weights. The x-axis denotes indices of assets, and the y-axis is each asset's average weight. For each asset, the center of an error bar denotes its portfolio mean (over 5651 trading days), and vertical lines denote its standard deviations: (g) Anticor; (h) PAMR; and (i) OLMAR.*

Li et al. 2011a) that the asset is important in all these approaches. Moreover, the increasing top five weights, which indicate more active exploitations, may lead to their increased performance. However, their volatilities also show that the subsets of assets are changing from day to day, which is inconvenient from the point of view of transaction costs. Anyway, such observations confirm that their pattern-matching process is improving and validate CORN's motivation.

The three mean reversion algorithms (Anticor, PAMR, and OLMAR) generally concentrate on the top five volatile stocks, as shown in Figure 13.12g through i and Table 13.6, while their orders may vary. Since Anticor, PAMR/CWMR, and OLMAR, in general, achieve the best performance on most other datasets, we also plot their average allocations in Table C.5,* in Appendix C. From the figure and tables, we can have several observations. First, similar to the pattern matching–based approaches, these algorithms have much higher volatilities than EG or ONS. However, different from the pattern matching–based algorithms, which only have higher volatilities on

---

*We ignore their corresponding figures, which are similar to Figure 13.12.

Table 13.6 *Top five (average) allocation weights of some strategies on NYSE (O)*

Asset #	6	23	9	26	20	Asset #	23	6	20	9	16
BCRP	0.28	0.25	0.20	0.18	0.09	EG	0.032	0.030	0.029	0.029	0.029
Asset #	8	35	2	3	22	Asset #	23	20	26	33	9
ONS	0.25	0.17	0.13	0.07	0.06	B^K	0.21	0.11	0.08	0.07	0.06
Asset #	23	20	9	6	26	Asset #	23	9	26	6	20
B^{NN}	0.21	0.15	0.08	0.08	0.08	CORN	0.38	0.09	0.09	0.09	0.08
Asset #	20	23	9	26	6	Asset #	23	20	9	26	6
Anticor	0.11	0.10	0.10	0.06	0.05	PAMR	0.19	0.11	0.11	0.08	0.06

top five weighted assets, the three algorithms also have much higher volatilities on other assets. Concerning their performance, it is possible that to achieve better performance, a portfolio has to be frequently rebalanced, not only on certain assets as the pattern matching–based algorithms do but also on all assets.

Second, most average weights of the state-of-the-art algorithms are assigned to the assets with the highest volatilities (highest Std values). It is common knowledge that high return is often associated with high risk,* while the reverse is not always true. That is, although a portfolio has to be rebalanced among volatile assets, such that the portfolio can gain profits from market volatility, high volatility cannot guarantee high profit. For example, on the NYSE (O) dataset, although Anticor and PAMR have the same top five average allocation pool, their performances are drastically different.

Third, PAMR, which systematically exploits the mean reversion property, rebalances more actively than Anticor, and OLMAR rebalances even more actively. Connecting the rebalance activities to their performance, we may conclude that even though both are based on the same principle, more active rebalance leads to better performance, as it can better exploit market volatility. PAMR's concentration on asset #23, which has the highest negative autocorrelation, sheds lights on the possible connection between mean reversion algorithms and the autocorrelation among assets (Lo and MacKinlay 1990; Conrad and Kaul 1998; Lo 2008). Moreover, from Table C.5, we can observe that most of the top average allocation weights of the mean reversion algorithms are assets with negative autocorrelations, except DJIA.

13.7   Summary

In this chapter, we empirically evaluated the four proposed algorithms. The empirical results clearly validate the effectiveness of the proposed algorithms. In terms of cumulative wealth, which is the main performance metric, our proposed algorithms sequentially beat the state-of-the-art algorithms. In terms of (volatility/drawdown) risk-adjusted return, the proposed algorithms achieve high risk-adjusted returns,

---

*Such a statement is true in traditional finance. However, in recent years, some arbitrage strategies, which can earn return without high risk, have emerged.

although they also have higher risk. The evaluations of parameter sensitivity show that the proposed algorithms are always robust to their parameters and have a wide range of satisfying choices such that they have good performance. The proposed algorithms are also scalable to two practical issues, that is, margin buying and transaction costs. Finally, although correlation-driven nonparametric learning (CORN) takes similar time as the state of the art, the three mean reversion costs significantly less time, which thus is suitable for practical large-scale applications, such as high-frequency trading.

In the future, we plan to study the sources of profits among online portfolio selection (OLPS). One way is to remove the possible "bid–ask bounce" by using a different methodology for computing the closing prices, such as averaging the prices of several transactions, which would reduce or even eliminate the bid–ask bounce. Moreover, incorporating other sources of information, such as volume information, is also possible to improve the proposed algorithms.

Chapter 14

Threats to Validity

Profitable real trading systems are complex systems, involving varying market scenarios. While the empirical results have demonstrated the effectiveness of the proposed strategies, there is still a long way to the production stage. In this chapter, we provides some arguments of various assumptions made during the trading model, back tests, and so on.

This chapter is organized as follows: Section 14.1 discusses the assumptions on the model, and Section 14.2 discusses the assumptions on the mean reversion principles. Section 14.3 discusses the proposed algorithms from a theoretical perspective. Section 14.4 validates the empirical studies. Finally, Section 14.5 summarizes this chapter and proposes some future directions.

14.1 On Model Assumptions

Any statement about such encouraging empirical results achieved by the proposed algorithms would be incomplete without acknowledging the simplified assumptions. To recall, we had made several assumptions regarding transaction cost, market liquidity, and market impact that would affect the algorithms' practical deployment.

The first assumption is that no transaction cost exists. In Section 13.4, we have already examined the effects of varying transaction costs, and the results show that the proposed algorithms can withstand moderate transaction costs in most cases. Currently, with the widespread adoption of electronic communication networks and multilateral trading facilities in financial markets, various online trading brokers charge very small transaction cost rates, especially for large institutional investors. They also use a flat rate,* which is based on the volume one reaches. Such measures can facilitate the portfolio managers to lower their transaction cost rates.

The second assumption is that the market is liquid and one can buy and sell any quantity at quoted prices. In practice, low market liquidity often means a large *bid–ask spread*—the gap between prices quoted for an immediate bid and an immediate ask. As a result, the execution of orders may incur a discrepancy between the prices

---

*For example, for US equities and options, E*Trade (https://global.etrade.com/gl/home, accessed on 16 March 2011.) charges only $9.99 for $50,000+ or 30+ stocks per quarter.

sent by algorithms and the prices actually executed. Moreover, stocks are often traded in multiples of *lots*, which is the standard trading unit containing a number of stock shares. In this situation, the quantity of the stocks may not be arbitrarily divisible. In our numerical evaluations, we have tried to minimize the effect of market liquidity by choosing the stocks that have large market capitalizations, which usually have small bid–ask spreads and discrepancies, and thus have high market liquidity.*

The third assumption is that a portfolio strategy would have no impact on the market, that is, the stock market will not be affected by any trading algorithms. In practice, the impact can be neglected if the market capitalization of a portfolio is not too large. However, as the experimental results show, the portfolio wealth generated by the proposed algorithms increases very fast, which would inevitably impact the markets. One simple way to handle this issue is to scale down the portfolio, as is done by many quantitative funds. Moreover, the development of sell-side algorithmic trading, which slices a big order into multiple smaller orders and schedules these orders to minimize their market impact, can significantly decrease the potential market impact of the proposed algorithms.

Here, we emphasize again that our current study assumes a "perfect market," which is consistent with existing studies in literature. It is important to note that even in such a perfect financial market, no algorithm has ever claimed such high performance, especially on the standard NYSE (O) dataset. Though past performance may not be a reliable indicator of future performance, such encouraging results do provide us confidence that the proposed algorithms may work well in future unseen markets.

14.2   On Mean Reversion Assumptions

Though the proposed mean reversion algorithms perform well on most datasets, we do not claim that they perform well on arbitrary portfolio pools. Note that passive–aggressive mean reversion (PAMR)/confidence-weighted mean reversion (CWMR) relies on the assumption that (single-period) mean reversion exists in a portfolio pool, that is, buying underperforming stocks in previous periods is profitable. Preceding experiments seem to show that, in most cases, such mean reversion does exist. However, it is still possible that this assumption fails to exist in certain cases, especially when portfolio components are incorrectly selected. PAMR/CWMR's performance on the DJIA dataset indicates that (single-period) mean reversion may not exist in the dataset. Although both are based on mean reversion, PAMR and Anticor are formulated with different time periods of mean reversion, which may be interpreted as meaning that Anticor achieves a good performance on DJIA. This also motivates the proposed online moving average reversion (OLMAR), which exploits multiple-period instead of single-period mean reversion. Thus, before investing in a real market, it is of crucial importance to ensure that the motivating mean reversion, either single period or multiple period, does exist among the portfolio pools. In academia, the mean reversion property in a single stock has been extensively studied

---

*However, we cannot say that we have removed or eliminated the impact of the bid–ask spread.

(Poterba and Summers 1988; Hillebrand 2003; Exley et al. 2004); one natural way is to calculate the sign of its autocorrelation (Poterba and Summers 1988). On the contrary, the mean reversion property among a portfolio lacks academic attention. Our motivation in CWMR (Table 10.1) provides a preliminary method to test single-period mean reversion. Different from the mean reversion in a single stock, mean reversion in a portfolio concerns not only the mean reversion in individual stocks but also the interactions among different stocks.

14.3   On Theoretical Analysis

In this book, our evaluations focus on empirical aspects of the strategies, which is unfair to some theoretically guaranteed methods, such as UP, EG, and ONS. Although the proposed four algorithms are not designed to asymptotically achieve the exponential growth of a specific experts, such as BCRP, it is better for us to explain the aspect of theoretical analysis, which is missing in our study.

On the one hand, we give no theoretical guarantee, or universal property, for the four proposed algorithms. In particular, we find it hard to prove the universal property for CORN, as it utilizes a correlation coefficient to select a similarity set. For the three mean reversion algorithms, since the mean reversion trading idea is counterintuitive, it is difficult to provide a traditional regret bound.* Although we cannot prove the traditional regret bound, the proposed algorithms do provide strong empirical evidence, which sequentially advances the state of the art.

On the other hand, it is possible to utilize certain meta-algorithms (Li et al. 2012, 2013; Li and Hoi 2012) that combine the proposed algorithms and some universal portfolio selection algorithms, such that the entire meta-system enjoys the universal property (Das and Banerjee 2011, Corollary 1). Meanwhile, such a meta-system can also benefit from the proposed algorithms and can produce significant high empirical performance. Note that even with a worst-case guarantee, some existing universal algorithms perform poorly on the datasets. Anyway, even though it is convenient to propose a universal meta-system, the original algorithms' theoretical aspects are still an open question and deserve further exploration.

14.4   On Back-Tests

Due to the unavailability of the intraday data and order books, we have conducted all the experiments based on public daily data, even though it may suffer from certain potential problems. One potential problem is that our algorithms may be earning "dealer's profits" in an uncontrolled and unfair way, or simply they are earning from the "bid–ask bounce" (Mcinish and Wood 1992; Porter 1992), which denotes a result of trades replacing the market maker's bid or ask quotes. This suspicion is compatible with the algorithms being contrarian strategies, such as PAMR, CWMR, and OLMAR. To eliminate this possibility, it would be good to try to eliminate the bid–ask bounce by replacing the market prices by the midpoint of the best bid and ask

---

*Borodin et al. (2004) failed to provide a regret bound for Anticor strategy, which passively exploits the mean reversion idea.

prices (Gosnell et al. 1996). However, calculating the midpoints of the best bid and ask prices requires access to the order book, which is usually private and not free, rather than simply the log of transactions. Another possibility would be to take into account only "sell-type" (or only "buy-type") transactions, meaning the transactions in response to market orders to sell, in which case the buying counterpart would be the one issuing a limit order. However, addressing the possibility also requires one to find out the order type (Keim and Madhavan 1995; Foucault et al. 2005) of each trade, which is usually not available to the public.

Back-tests in historical markets may suffer from "data-snooping bias" issues, one of which is the dataset selection issue. On the one hand, we selected four datasets, the NYSE (O), TSE, SP500, and DJIA datasets, based on previous studies without consideration to the proposed approaches. On the other hand, we developed the proposed algorithms solely based on the NYSE (O) dataset, while the other five datasets (NYSE (N), TSE, SP500, MSCI, and DJIA) were obtained after the algorithms were fully developed. However, even though we are cautious about the dataset selection issue, it may still appear in the experiments, especially for the datasets with a relatively long history, that is, NYSE (O) and NYSE (N). The NYSE (O) dataset, pioneered by Cover (1991) and followed by other researchers, is a "standard" dataset in the online portfolio selection community. Since it contains 36 large-cap NYSE stocks that survived for 22 years, it suffers from extreme survival bias. Nevertheless, it still has the merit to compare different algorithms as done in all previous studies. The NYSE (N) dataset, as a continuation of NYSE (O), contains 23 assets that survived from the previous 36 stocks for another 25 years. Therefore, it becomes even worse than its precedent in terms of survival bias. In summary, even though the empirical results on these datasets clearly show the effectiveness of the proposed algorithms, one cannot make claims without noticing the deficiencies of these datasets.

Another common bias is the asset selection issue. Four of the six datasets (the NYSE (O), TSE, SP500, and DJIA) are collected by others, and to the best of our knowledge, their assets are mainly the largest blue chip stocks in their respective markets. As a continuation of NYSE (O), we self-collected NYSE (N), which again contains several of the largest survival stocks in NYSE (O). The remaining dataset (MSCI)* is chosen according to the world indices. In summary, we try to avoid the asset selection bias via arbitrarily choosing some representative stocks in their respective markets, which usually have large capitalization and high liquidity and thus reduce the market impact caused by any proposed portfolio strategies.

Moreover, there are some critics regarding the datasets' liquidity issue, which assumes that the assets are available in unbounded quantities for buying or selling at any given trading period. In Table 13.1, we observe cumulative of 10^{13} or more, and there are assets with capitalization less than 10^{10}; then, obviously, the liquidity assumption is not fulfilled. In NYSE (O), there are many such assets, and even in NYSE (N) there are four such assets: SHERW, KODAK, COMME, and KINAR. The most "dangerous" asset is KINAR, identified as asset #23 in Table 13.5, where there

---

*In fact, we collected this dataset following Li et al. (2012)'s review comments, which means the dataset does not exist before its third-round submission.

are no data on its capitalization, but certainly it is a very small asset. One remedy is to only consider the remaining 19 assets out of the 23 in the experiments, as done by Györfi et al. (2012, Chapter 2).

Finally, following existing model assumptions and experimental setting, we do not consider the low-quality assets, such as the bankrupt and penny stocks. The bankrupt stock data are difficult to acquire; thus, we cannot observe their behaviors and predict the behaviors of the proposed algorithms. In reality, the bankruptcy situation rarely happens for blue chip stocks because typically a bankrupt stock would be removed from the list of blue chip stocks before it actually goes into bankruptcy. The penny stocks lack sufficient liquidity to support the trading frequency required for our current research. Besides, one could also explore many practical strategies to exclude such low-quality stocks from the asset pool at some early stage, such as technical and fundamental analysis.

14.5 Summary

This chapter argued some assumptions in our models and back-tests, which will be faced by various empirical research in trading strategies. When back-testing a strategy, researchers should be aware of these assumptions and thus can take measures to weaken their impacts on the profits in real trading.

Part V

Conclusion

Chapter 15

Conclusions

If there's one thing I learned in prison
it's that money is not the prime commodity in our lives...
time is.
– Wall Street 2: Money Never Sleeps

15.1 Conclusions

This book aims to advance the state of the art in online portfolio selection (OLPS). Here, our objective is to achieve better performance on real markets. The main principles we adopted are the principles of pattern matching and mean reversion.

For the principle of pattern analysis, we try to locate similar patterns from the historical market and construct optimal portfolios based on these patterns. Observing that existing pattern matching–based approaches often adopt Euclidean distance to measure the similarity between two patterns, we find that Euclidean distance ignores their linear similarity and whole-market movements. Thus, we proposed to measure the similarity via a correlation coefficient, which considers both ignored aspects, and designed the CORrelation-driven Nonparametric learning (CORN) approach for OLPS. The proposed CORN performs much better than existing pattern matching–based strategies, which validates its motivations.

For mean reversion, we directly output portfolios based on the principle, which assumes that the price trends will revert to their previous trends. Firstly, we proposed to exploit the principle via passive–aggressive learning, resulting in the passive–aggressive mean reversion (PAMR). In particular, PAMR tries to obtain portfolios that perform worse than a threshold on the last price relatives, and also close to the last portfolio. PAMR's formulation is clear to understand, and its closed-form solutions reflect the mean reversion principle. PAMR can achieve the best empirical performance on most datasets at the time.

Observing that most existing algorithms only exploit the first-order information of a portfolio, we proposed to exploit the second-order information and the mean reversion property via confidence-weighted learning, resulting in a new family of strategies called confidence-weighted mean reversion (CWMR). It models the portfolio as a Gaussian distribution and sequentially updates the distribution similar to

PAMR, but exploits both first-order and second-order information. CWMR's closed-form updates effectively trade off between the first- and second-order information of a portfolio. Empirically, it generally outperforms other existing strategies, including CORN and PAMR, on most datasets.

Analyzing the existing algorithms via Kelly's framework, we find that the above two mean reversion algorithms follow the assumption of single-period mean reversion, which leads to performance degradation on certain datasets. To handle the degradation and to further exploit the market, we proposed two forms of multiperiod mean reversion, both of which are based on a moving average, and the online moving average reversion (OLMAR), which is more robust than PAMR. Empirically, OLMAR is currently the best strategy, beating all existing algorithms, including CORN, PAMR, and CWMR.

We conducted an extensive set of empirical evaluations, in which the results clearly validate the effectiveness of the proposed algorithms. In particular, the proposed algorithms sequentially advance the state of the art in terms of cumulative return, which is the main performance metric of our studies. Besides, they are fairly robust to their parameter settings, and most of the algorithms are generally computationally efficient and are thus suitable for real-life large-scale environments.

15.2 Future Directions

15.2.1 On Existing Work

We have presented a family of algorithms for OLPS that represent the state of the art of OLPS in academia. However, there is still room to further improve the existing algorithms.

First of all, we calculate CORN's correlation coefficient using a univariate correlation by concatenating each market window to a column vector, during which we may lose useful structural information. Thus, it is possible to improve its performance using a multivariate correlation coefficient on the matrices such that the information is retained. Since information is crucial for a portfolio selection task, we can construct more effective portfolios. Another potential improvement is to redesign the CORN algorithm when transaction costs exist, since the cumulative wealth achieved by CORN decreases exponentially with increasing transaction costs. One possible solution is to add a regularization term to the portfolio maximization step, such that we can maximize expected return and meanwhile constrain expected turnovers.

Moreover, currently we only consider the similarity between two market windows with the same length and same interval; however, locating patterns with varying timing is also attractive in the pattern matching–based approaches. Dynamic time warping (DTW) (Rabiner and Levinson 1981; Sakoe and Chiba 1990; Keogh 2002) is a dynamic programming approach proposed to recognize humans' spoken words, which vary a lot in timing and pronunciation. Since similarity among time series also varies in time, DTW has been successfully applied to find patterns in time series (Berndt and Clifford 1994; Yi et al. 1998; Keogh and Pazzani 2000; Rakthanmanon et al. 2012). In finding patterns in time series, DTW's basic idea is to stretch or compress the time axis, such that the distance between time series and a template is minimized. In this way, one template can match various time series with varying time. Since the pattern matching–based approaches try to locate similar patterns among the historical time series, DTW may be directly applied to locate time-warping patterns. To the best of our knowledge, no existing algorithm has ever considered the time compression or stretch of market windows. However, one associated problem with DTW in practice is its expensive time cost (Wang et al. 2013).

Second, although the three mean reversion algorithms outperform the previous algorithms on most datasets, they may fail in certain cases. For example, if the market contains one stock that drops continuously and significantly over some periods, then all the three algorithms may fail. Such scenarios may simulate a financial crisis when all stocks continuously drop for a certain period, or the case that one stock is hit by a series of bad news, and its price slashes continuously over a certain period. In fact, for such cases, most current good-performing algorithms, including Anticor and all pattern matching–based algorithms, will underperform the naive market strategy. In our experiments, we choose blue chip stocks, by assuming that they would not drop continuously for a long period. However, it is always possible that the "black swan" (Taleb 2008) may exist in the financial market and hit your portfolio. Such a hypothetical market sequence poses a serious challenge: How to achieve a reasonable performance in such a market?

Third, in Section 9.4, we briefly connect PAMR's update and the general form of return-based contrarian strategies. Besides PAMR, we find that GP's update in Section 4.2 also coincides with the general form of return-based momentum strategies. This finding is not abnormal as both adopt similar learning techniques but contrary trading ideas, that is, contrarian for PAMR and momentum for GP. Based on such a connection, we can anatomize a portfolio's expected return and find its sources of profit. Although pure trading strategies have been anatomized for a long time, OLPS algorithms have yet to be anatomized. Such an analysis will help understand the behaviors of OLPS algorithms, which is largely unknown in current research. The resulting empirical observations can help validate the analysis, as is done in previous studies (Lo and MacKinlay 1990; Conrad and Kaul 1998; Lo 2008).

Fourth, as empirically analyzed in Section 13.6 and discussed in Section 14, different asset pools are suitable for different algorithms. Also, as there exist thousands of assets in the financial markets, it would be more computationally efficient to select a subset of assets. This poses an open challenge, that is, how to prepare a proper asset pool for specific OLPS algorithms. This challenge is similar to the task of feature selection, and we plan to explore general methods to automatically select effective subsets of assets.

Finally, although all the proposed algorithms significantly outperform the state of the art, one important aspect missing in current study is their theoretical guarantees. As discussed in Section 14.3, there exist several explanations and measures to handle the issue. Nevertheless, the theoretical guarantees, either the universal property or others, are still missing for the proposed algorithms. In future, we plan to present some nontrivial theoretical guarantees for the proposed algorithms, such that we can be more confident regarding the algorithms' practical applicability.

15.2.2   On Practical Issues

To improve the practicability of OLPS algorithms, one should also tackle the practical issues in real markets, so as to relax the ideal assumptions of the market model and provide possible applications.

A crucial practical issue is transaction cost (Kissell et al. 2003), which has been addressed in some literatures (Blum and Kalai 1999; Iyengar 2005; Lobo et al. 2007; Györfi and Vajda 2008; Kozat and Singer 2008, 2010). Among them, a research direction is to investigate a simple linear model with a proportional transaction cost (Blum and Kalai 1999; Borodin et al. 2004; Györfi and Vajda 2008), which is easier to handle than nonlinear models (Takano and Gotoh 2011). Different from existing solutions, which mainly trade off between the expected return and expected transaction costs, one can solve this by the regularization methods (Tibshirani 1996) to resolve ill-posed problems and minimize portfolio turnover at a reasonable expected return.

In addition to addressing explicit transaction costs, another interesting future direction is to ignore commission fees and taxes but consider them as part of the bid–ask spread, which allows trading strategies to issue not only market but also limit orders. This would allow for negative transaction costs,* which become profit of the dealer/algorithm.

Another practical issue in the task is to incorporate useful side information, such as experts' opinions, firms' fundamental information, and technical indicators. As shown in existing studies (Cover and Ordentlich 1996; Fagiuoli et al. 2007), such side information may facilitate the portfolio selection task. Moreover, other state-of-the-art research (Zhang and Skiena 2008, 2010) focuses on learning real-time news to facilitate a single stock trading, while portfolio trading is still new. Most existing works manually evaluate the side information as integers, and one potential work is to sequentially learn appropriate integers with the side information. The learned integers can help us to handle the side information and thus improve the performance of the proposed approaches.

15.2.3   Learning for Index Tracking

The literature (Meade and Salkin 1989, 1990) shows that most active management funds focusing on absolute return usually do not outperform the corresponding market index. Thus, index mutual funds, which try to track the market index, have emerged in recent decades. Briefly speaking, index tracking is passive portfolio management aiming to construct a portfolio that tracks one real or virtual index using stock shares of as few companies as possible. Currently, some computational intelligence approaches (Gilli and Këllezi 2002; Beasley et al. 2003; Coleman et al. 2006; Maringer 2008; Canakgoz and Beasley 2009) have been proposed to tackle this task. We also notice the development of lasso techniques (Tibshirani 1996; Fu 1998; Zou 2006; Bach 2008; Yang et al. 2010; Zhou et al. 2010), which mainly control the sparsity of a decision variable. Thus, it is possible to handle the index-tracking problem via the lasso techniques (McWilliams and Montana 2010). On the one hand, it can

---

*By issuing limit orders, the algorithm would play the traditional role of a dealer.

ensure that the required tracking error is minimized on the constraint that the number of chosen companies is less than a predefined number. On the other hand, we can stratify the companies into groups and achieve the sparsity among groups or within a group. The main challenge to learn an index-tracking portfolio using lasso techniques is the contradiction between the simplex constraint and the lasso constraint, as the former deactivates the lasso constraint. Brodie et al. (2009) solved this challenge in the traditional mean variance model, while how to tackle this challenge in the context of online index tracking can be further investigated.

OLPS: A Toolbox for Online Portfolio Selection

This appendix presents an open-source software toolbox for online portfolio selection (OLPS), which implements a collection of classical and state-of-the-art OLPS strategies powered by state-of-the-art machine learning techniques.

OLPS aims to sequentially allocate capital among a set of assets to maximize long-term return. In recent years, a variety of machine learning algorithms have been proposed to address this challenging problem, but there is no comprehensive open-source toolbox available due to various reasons. This may have significantly hindered the progresses of research and development of new techniques in this field.

OLPS was designed and developed to facilitate the investigation and development of new methods and enable performance benchmarking and comparisons of different existing strategies. OLPS is an open-source project implemented in MATLAB® and is compatible with Octave. The software toolbox has been released under Apache License (version 2.0), and it is freely available at: https://github.com/OLPS/

A.1   Introduction

A.1.1   Target Task

In this section, we briefly formulate the OLPS model, which will be used in our model. Suppose we have a finite number of $m \geq 2$ investment assets, over which an investor can invest for a finite number of $n \geq 1$ periods.

At the t-th period, $t = 1, \ldots, n$, the asset (close) prices are represented by a vector $\mathbf{p}_t \in \mathbb{R}_+^m$, and each element $p_{t,i}, i = 1, \ldots, m$ represents the close price of asset i. Their price changes are represented by a *price relative vector* $\mathbf{x}_t \in \mathbb{R}_+^m$, each component of which denotes the ratio of the t-th close price to the last close price, that is, $x_{t,i} = \frac{p_{t,i}}{p_{t-1,i}}$. Thus, an investment in asset i throughout period t changes by a factor of $x_{t,i}$. Let us denote by $\mathbf{x}_1^n = \{\mathbf{x}_1, \ldots, \mathbf{x}_n\}$ a sequence of price relative vectors for n periods, and $\mathbf{x}_s^e = \{\mathbf{x}_s, \ldots, \mathbf{x}_e\}, 1 \leq s < e \leq n$ as a market window.

An investment in the market for the t-th period is specified by a *portfolio vector* $\mathbf{b}_t = (b_{t,1}, \ldots, b_{t,m})$, where $b_{t,i}, i = 1, \ldots, m$, represents the proportion of wealth invested in asset i at the beginning of the t-th period. Typically, a portfolio is self-financed and no margin/short sale is allowed; therefore, each entry of a portfolio is nonnegative and adds up to one, that is, $\mathbf{b}_t \in \Delta_m$, where $\Delta_m = \{\mathbf{b}_t : \mathbf{b}_t \succeq 0, \sum_{i=1}^m b_{t,i} = 1\}$.* The investment procedure is represented by a *portfolio strategy*, that is, $\mathbf{b}_1 = \left(\frac{1}{m}, \ldots, \frac{1}{m}\right)$ and the following sequence of mappings:

$$\mathbf{b}_t : \mathbb{R}_+^{m(t-1)} \rightarrow \Delta_m, t = 2, 3, \ldots,$$

where $\mathbf{b}_t = \mathbf{b}_t(\mathbf{x}_1^{t-1})$ is the portfolio determined at the beginning of the t-th period upon observing past market behaviors. We denote by $\mathbf{b}_1^n = \{\mathbf{b}_1, \ldots, \mathbf{b}_n\}$ the strategy for n periods, which is the output of an OLPS strategy.

At the t-th period, a portfolio $\mathbf{b}_t$ produces a *portfolio period return* s_t, that is, the wealth changes by a factor of $s_t = \mathbf{b}_t^\top \mathbf{x}_t = \sum_{i=1}^m b_{t,i} x_{t,i}$. Since we reinvest and adopt relative prices, the wealth would change multiplicatively. Thus, after n periods, a portfolio strategy $\mathbf{b}_1^n$ will produce a *portfolio cumulative wealth* of S_n, which changes the initial wealth by a factor of $\prod_{t=1}^n \mathbf{b}_t^\top \mathbf{x}_t$:

$$S_n(\mathbf{b}_1^n, \mathbf{x}_1^n) = S_0 \prod_{t=1}^n \mathbf{b}_t^\top \mathbf{x}_t,$$

where S_0 denotes the initial wealth and is set to \$1 for convenience.

Algorithm A.1: Online portfolio selection.

Input: $\mathbf{x}_1^n$: Historical market price relative sequence
Output: S_n: Final cumulative wealth

Initialize $S_0 = 1, \mathbf{b}_1 = \left(\frac{1}{m}, \ldots, \frac{1}{m}\right)$;
for $t = 1, 2, \ldots, n$ **do**
  Portfolio manager learns a portfolio $\mathbf{b}_t$;
  Market reveals a price relative vector $\mathbf{x}_t$;
  Portfolio incurs period return $s_t = \mathbf{b}_t^\top \mathbf{x}_t$ and updates cumulative return $S_t = S_{t-1} \times (\mathbf{b}_t^\top \mathbf{x}_t)$;
  Portfolio manager updates his/her decision rules;
end

We present the framework of the above task in Algorithm A.1. In this task, a portfolio manager's goal is to produce a portfolio strategy ($\mathbf{b}_1^n$) upon the market price relatives ($\mathbf{x}_1^n$), aiming to achieve certain targets. He or she computes the portfolios in a sequential manner. At each period t, the manager has access to the sequence of past price relative vectors $\mathbf{x}_1^{t-1}$. He or she then computes a new portfolio $\mathbf{b}_t$ for next price relative vector $\mathbf{x}_t$, where the decision criterion varies among different managers.

---

*$\succeq 0$ denotes that each element of the vector is nonnegative.

Then the manager will rebalance to the new portfolio via buying and selling the underlying stocks. At the end of a trading period, the market will reveal $\mathbf{x}_t$. The resulting portfolio $\mathbf{b}_t$ is scored based on portfolio period return s_t. This procedure is repeated until the end, and the portfolio strategy is finally scored by the portfolio cumulative wealth S_n.

Note that we have made several general and common nontrivial assumptions in the above model:

1. Transaction cost: no explicit or implicit transaction costs* exist.
2. Market liquidity: one can buy and sell the required amount, even fractional, at the last close price of any given trading period.
3. Market impact: any portfolio selection strategy shall not influence the market or any other stocks' prices.

All the implemented strategies follow the same architecture in Algorithm A.1, and they are called at Line 3.

A.1.2   Installation

A.1.2.1   Supported Platforms

OLPS is based on MATLAB (both 32- and 64-bit) and Octave (except the Graphical User Interface [GUI] Part); thus, it is supported on 32- and 64-bit versions of Linux, Mac OS, and Windows. The first version of OLPS is developed and tested on MATLAB 2009a, while the latest version of OLPS is tested on MATLAB 2013a.

A.1.2.2   Installation Instructions

Installation of the toolbox consists of two steps:

1. Retrieve the latest version of OLPS from the project website. The package can be downloaded as either a .zip file or a .tar.gz file.
2. Unpack the package to any folder. The root directory is named "OLPS."

Then the toolbox is available in the folder. Note that the directory structure of the toolbox is predefined, which decides the running datasets and logs.

A.1.2.3   Folders and Paths

The toolbox consists of five folders in relative path: "/Strategy," "/Data," "/GUI," "/Log," and "/Documentation." The folder "/Strategy" consists of the core strategies for online portfolios selection, which will be introduced in Section A.3. The folder also consists of the commands used in the Command Line Interface (CLI), which will be introduced in Section A.2.2. The folder "/Data" includes some popular datasets in forms of .mat, which will be detailed in Section A.1.4. The folder "/GUI" includes the files to run the Graphical User Interface, which will be detailed in Section A.2.1. The folder "/Log" stores the experimental details of a strategy on a dataset, which will be

---

*Explicit costs include commissions, taxes, stamp duties, and fees. Implicit costs include the bid–ask spread, opportunity costs, and slippage costs.

generated after the simulation process. The folder "/Documentation" contains some documentations of the toolbox, including one summary paper and one comprehensive documentation of the toolbox.

A.1.3   Implemented Strategies

Table A.1 illustrates all implemented strategies in the toolbox.

Table A.1 *All implemented strategies in the toolbox*

Categories	Strategies	Sections	Strategy Names
Benchmarks	Uniform Buy and Hold	A.3.1.1	ubah
	Best Stock	A.3.1.2	best
	Uniform Constant Rebalanced Portfolios	A.3.1.3	ucrp
	Best Constant Rebalanced Portfolios	A.3.1.4	bcrp
Follow the Winner	Universal Portfolios	A.3.2.1	up
	Exponential Gradient	A.3.2.2	eg
	Online Newton Step	A.3.2.3	ons
Follow the Loser	Anticorrelation	A.3.3.1	anticor/anticor_anticor
	Passive–Aggressive Mean Reversion	A.3.3.2	pamr/pamr_1/pamr_2
	Confidence-Weighted Mean Reversion	A.3.3.3	cwmr_var/cwmr_stdev
	Online Moving Average Reversion	A.3.3.4	olmar1/olmar2
Pattern Matching	Nonparametric Kernel-Based Log-Optimal	A.3.4.1	bk
	Nonparametric Nearest Neighbor Log-Optimal	A.3.4.2	bnn
	Correlation-Driven Nonparametric Learning	A.3.4.3	corn/cornu/cornk
Others	M0		m0
	T0		t0

Table A.2 *All included datasets in the toolbox*

File Names (.mat)	Dataset	Region	Time Frame	# Periods	# Assets
nyse-o	NYSE (O)	US	07/03/1962–12/31/1984	5651	36
nyse-n	NYSE (N)	US	01/01/1985–06/30/2010	6431	23
tse	TSE	CA	01/04/1994–12/31/1998	1259	88
sp500	SP500	US	01/02/1998–01/31/2003	1276	25
msci	MSCI	Global	04/01/2006–03/31/2010	1043	24
djia	DJIA	US	01/14/2001–01/14/2003	507	30

A.1.4 Included Datasets

As shown in Table A.2, six main datasets are widely used for the OLPS task. We do not include the high-frequency datasets (Li et al. 2013) as they are private. Other variants, such as the revered datasets (Borodin et al. 2004) and margin datasets (Helmbold et al. 1998), which users may generate themselves, will not be provided in the toolbox.

A.1.5 Quick Start

To quick start the OLPS toolbox, we provide two fast-entry files. One is GUI_start.m, which starts the GUI. The other is CLI_demo.m, which provides fast executions of all strategies one by one in the command line. All the parameters used in the file are set according to their original studies, respectively.

A.2 Framework and Interfaces

In this toolbox, we provide two interfaces to call the implemented strategies, that is, Graphical User Interface (GUI) and Command Line Interface (CLI). The framework can be easily extended to include new algorithms and datasets.

A.2.1 Graphical User Interface

In the GUI, users will call the implemented algorithms via interaction with the GUI. We provide a menu-driven interface for the user to select datasets and algorithms, and input the desired arguments. After providing the inputs and hitting the start button, the algorithm(s) execute. Upon completion, the results and relevant graphs are displayed.

A.2.1.1 Getting Started

To start the GUI, we type the following command in MATLAB:

```
>> OLPS_gui
```

After executing the above command, the *Trading Manager* starts. As shown in Figure A.1, the opening window has five buttons. The *About* and *Exit* buttons are self-explanatory. The other three are the main functional buttons. The *Algorithm*

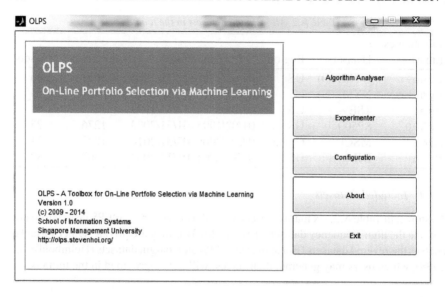

Figure A.1 *Starting the Trading Manager.*

Analyser button will start a new window, in which the user can run a single algorithm and analyze its performance relative to the basic benchmarks. The *Experimenter* button is used for selecting multiple algorithms and comparing their performances. The *Configuration* button is used to add or delete algorithms and datasets that can be used by the toolbox.

A.2.1.2   Algorithm Analyser

On pressing the *Algorithm Analyser* button, a new window opens that will be used for running and analyzing an algorithm. Figure A.2 depicts the Algorithm Analyser running the *Online Moving Average Reversion* on the *S&P500* dataset. There are drop-down menus for selecting the algorithm and the dataset. The input parameter fields will dynamically change depending on the inputs the algorithm requires (default parameters have been provided). When a particular dataset is selected, some preliminary performance details of the algorithm are displayed. There are three types of preliminary results displayed. *Basic Benchmarks* displays the cumulative returns for four simple algorithms—*Uniform Buy and Hold, Uniform Constant Rebalanced Portfolio, Best Stock* in hindsight, and *Best Constant Rebalanced Portfolio* (BCRP). For more details on these algorithms, refer to Section A.3. The *Returns Distribution* shows the annualized mean return and standard deviation of each asset in the dataset. The *All Assets* option shows the performance graph of cumulative returns of all the assets in the dataset.

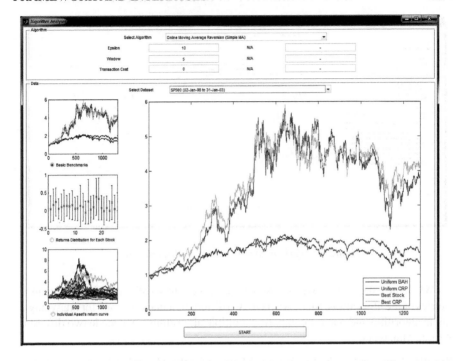

Figure A.2 *Various components of the Algorithm Analyser.*

A.2.1.3 Experimenter

When devising trading strategies, we usually want to compare the performance of these strategies relative to each other. For this purpose, we provide the *Experimenter*. On pressing the Experimenter button, a new window opens that will offer us the platform for comparing different strategies. First, the dataset is selected. From the list of algorithms, a subset can be selected to be executed. Among the selected algorithms, the input parameters have to be provided and saved (default values are already there). Figure A.3 gives an example of comparing six strategies on the *MSCI World Index* dataset. The six algorithms being compared are *Uniform Buy & Hold, Uniform Constant Rebalanced Portfolio, Best Constant Rebalanced Portfolio, Passive–Aggressive Mean Reversion, Confidence Weighted Mean Reversion*, and *Online Moving Average Reversion*.

A.2.1.4 Results Manager

After hitting the *Start* button in the *Algorithm Analyser* or *Experimenter*, the execution starts. In the Algorithm Analyzer, a progress bar indicates the execution of a single algorithm. In the case of the Experimenter, there are two progress bars. One indicates the number of algorithms executed (along with which algorithm is being executed currently) and the other shows the status of completion of that individual algorithm.

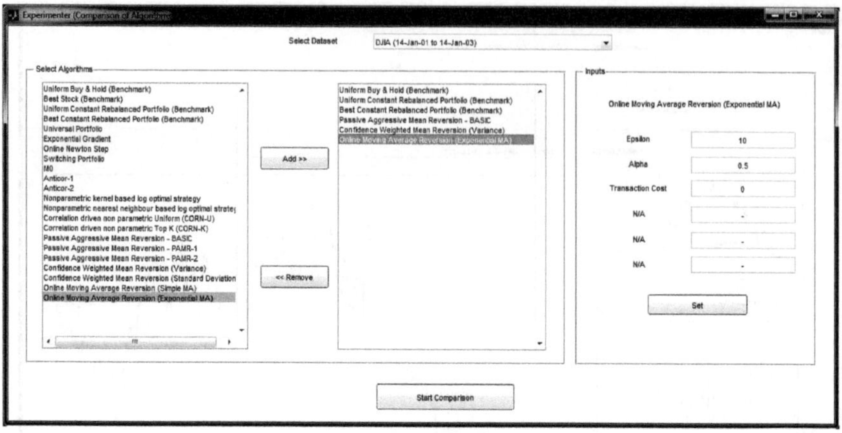

Figure A.3 *Various components of the Experimenter.*

When the execution is over, the *Results Manager* shows all the basic performance metrics of the algorithms. Since we have two different managers—one for analyzing a single algorithm and one for comparing multiple algorithms—we made two different Results Managers.

Results Manager 1   The first Results Manager for the *Algorithm Analyzer* is shown in Figure A.4. The table in the window quantifies the results of the algorithm as compared to the basic benchmarks. The numbers from this table can directly be copied and pasted. There is a large graph space that displays the information on a particular attribute selected in the left column.

Returns   It contains information about the daily performance of the algorithm. The user can choose to view the cumulative returns and the daily returns. The option of a log (base 10) plot is provided for easier visualization when the difference in performance of the algorithm and the benchmarks is significantly high.

Risk Analysis   There are five metrics to evaluate the risk and risk-adjusted returns of the algorithm. They are the *Sharpe Ratio*, *Calmar Ratio*, *Sortino Ratio*, *Value at Risk*, and *Maximum Draw down*. An input box called *Window* is provided next to each metric. The purpose of the window is to analyze the consistency of the algorithm, instead of just the final result. For example, entering 252 in the Sharpe Ratio Window will plot a graph of the Sharpe Ratio of the algorithm for time period $t - 252$ to t, for all t. When the window size is large such that t is less than the window size, then the computation starts from $t = 1$. The risk metrics are assumed to be zero for the first 50 time periods. This has been done to avoid extreme values due to lack of data in the initial periods.

Portfolio Analysis   The *Portfolio Allocation* shows the distribution of wealth allocated to each asset by the algorithm. The *Step by Step* helps us look at the portfolio allocation for any particular given day. Lastly, we have a portfolio *Animation* that accepts an input called *Window*. Visualizing portfolio changes based on daily frequency can be overwhelming and difficult to interpret, especially when the daily

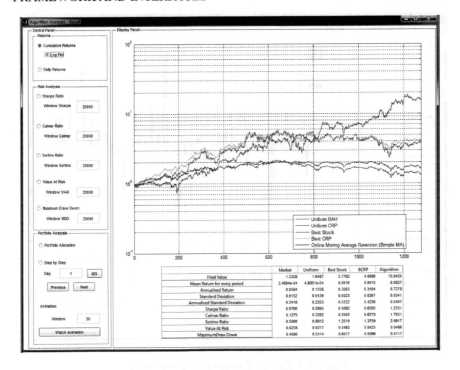

Figure A.4 *Results Manager for Algorithm Analyser.*

portfolio changes are significant. Instead, we allow the user to choose a moving average portfolio of the last *Window* number of days. This results in a smoother change
of the portfolio allocation.

Result Manager 2   The second Results Manager is very similar to the first manager,
except that it is designed for the *Experimenter*. The table in the window quantifies
the performance of the algorithms relative to each other. Like the first manager, this
manager also has three sections. A preview of this manager can be seen in Figure A.5.

   Returns   The daily returns across the entire time period of the dataset for all the
algorithms can be overwhelming to view. A time period can be selected, and the daily
performance of the algorithms is displayed for only that time period.

   Risk Analysis   This section is almost identical to that of the first Results Manager.
The only difference is that here the metrics are evaluated for every algorithm and
displayed together.

   Portfolio Analysis   This shows the distribution of portfolio allocation for all the
algorithms.

A.2.1.5   Configuration Manager

Here, we describe how to add or delete new algorithms and datasets via the
Configuration Manager, as shown in Figure A.6.

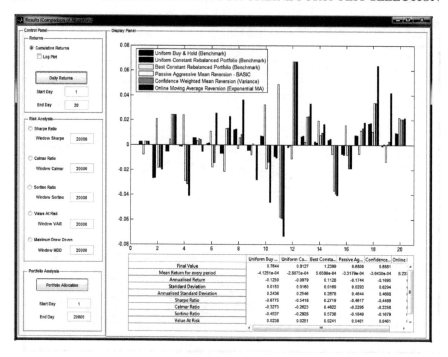

Figure A.5 *Results Manager for Experimenter.*

New Strategy   A template ("template.m") has been provided in the Strategy folder that is based on the general framework for OLPS (as described in Algorithm A.1). The user should enter his code to learn the new portfolio within the specified region of the loop. Without any changes to the code, the template will behave as a *Uniform Constant Rebalanced Portfolio* strategy, owing to the fact that we start with a uniform portfolio and never update it. All new strategies coded must remain in the Strategy folder. Once the files are created in the folders, the configuration should be changed using the Configuration Manager GUI, which controls the loading of algorithms and datasets into the Trading Manager.

New Dataset   A dataset is in the form of price relative vectors of various assets. The t-th row represents the price relative of all the assets at time t. The user just has to save the new price relative matrix in the Data folder. Data of different frequencies can be used as well. All the datasets provided in the toolbox are of daily frequency. Once the files are created in the folders, the configuration should be changed using the Configuration Manager GUI, which controls the loading of algorithms and datasets into the Trading Manager.

Configuration   The configuration determines the algorithms and datasets being used in the toolbox. Within the *config* folder, there is a file called *config*. This is the active configuration, which means the toolbox uses this file to determine which algorithms and datasets would be preloaded. There is another file *config_default*, which is the

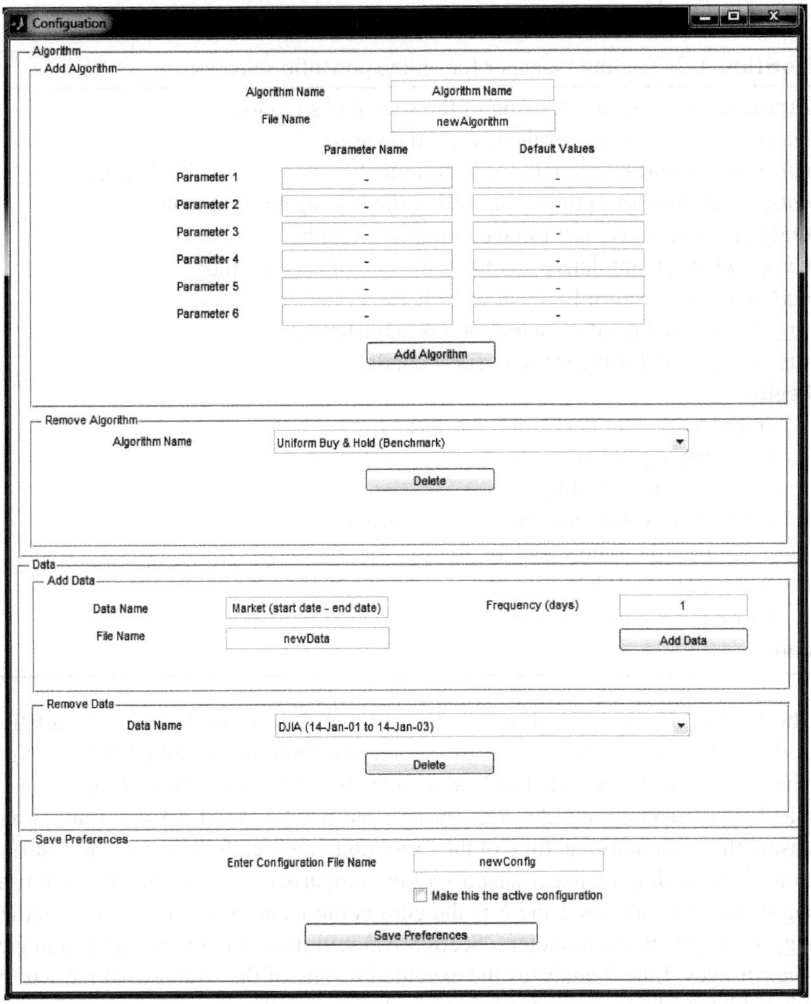

Figure A.6 *Configuration Manager.*

configuration provided by the toolbox. Initially, the content of the default and the active configuration are the same. A new configuration can be created by clicking on the *Configuration* button in the start window. It automatically loads the active configuration, to which the user can add or delete new algorithms or datasets.

A.2.2 Command Line Interface

In the CLI, users can run algorithms by calling the commands. In particular, we provide a meta-function named *manager*, which is responsible for preprocessing (such as initializing datasets and variables, etc.), calling specified strategies, and postprocessing (such as analyzing and outputting the results, etc.).

A.2.2.1   Trading Manager

Algorithm A.2: Trading manager for online portfolio selection.

Input: *strategy_name*: A string of the specified strategy;
dataset_name: A string of the specified dataset;
varargins: A variable-length input argument list for the specified strategy;
opts: A variable for options controlling the trading environment.
Output: *cumulative_ret*: Final cumulative wealth;
Cumprod_ret: Cumulative wealth at the end of each period;
daily ret: Daily return for each period;
ra_et: Analyzed results, including risk-adjusted returns;
run_time: Time for the strategy (in seconds).
begin

> Initialize market data from *dataset*;
> Open the log file and mat file;
> Start the time variables;
> Call *strategy* with parameters in *varargins*;
> Terminate the time variables;
> Analyze the results;
> Close the log file and mat file;

end

The *Trading Manager*, as shown in Algorithm A.2, controls the whole simulation of OLPS. At the start (Line 2), it loads market data from the specified dataset. Note that this can be easily extended to load data from real brokers. Then, Lines 3 and 8 open and close two logging files, one for text and one for .MAT format. Lines 4 and 6 measure the computational time of the execution of a specified strategy. Measuring the time in the trading manager ensures a fair comparison of the computational time among different strategies. Line 5 is the core component, which calls the specified strategy with specified parameters. Section A.3 will illustrate all included strategies and their usages. Line 7 analyzes the executed results of the strategy, which will be introduced later. The "manager.m" usage is shown as follows.

Usage

```
function [cum_ret, cumprod_ret, daily_ret, ra_ret,
        run_time]...
   = manager(strategy name, dataset name, varargins,
     opts);
```

- cum_ret: cumulative return;
- cumprod_ret: a vector of cumulative returns at the end of every trading day;
- daily_ret: a vector of daily returns at the end of every trading day;
- ra_ret: analyzed result;
- run_time: computational time of the core strategy (excluding the manager routine);

- strategy_name: the name of the strategy (all implemented strategies' names are listed in the fourth column of Table A.1);
- dataset_name: the name of the dataset;
- varargins: variable-length input argument list; and
- opts: options for behavioral control.

Example   This example calls the ubah (Uniform Buy and Hold, or commonly referred to as the market strategy) strategy on the "NYSE (O)" dataset.

```
[cum_ret, cumprod_ret, daily_ret, ra_ret, run_time]...
    = manager('ubah', 'nyse-o', {0}, opts);
```

To facilitate the debugging of trading strategies, we also use controlling variables to control the trading environment. In particular, the last parameter *opts* in the above example contains the controlling variables. As shown in Table A.3, it consists of five controlling variables.

The *Results Manager* analyzes the results and returns an array containing the basic statistics, the Sharpe ratio and Calmar ratio, and their related statistics. Details about the returned statistics are described in Table A.4.

Usage

```
function [ra_ret] ...
    = ra_result_analyze(fid, data, cum_ret, cumprod_ret,
                        daily_ret, opts);
```

Adding Your Own Strategy or Data   Adding new strategies and datasets in the CLI mode is similar to that in the GUI mode. Adding the strategy involves replacing the portfolio update component of the algorithms, and adding a dataset involves storing the market matrix and placing the files in the data folder.

Table A.3 *Controlling variables*

Variables	Descriptions	Possible Values	Explanation for Values
opts.quiet_mode	display debug info?	0 or 1	No or Yes
opts.display_interval	display info time interval?	Any number (e.g., 500)	Display every 500 periods
opts.log_record	record the .log file?	0 or 1	No or Yes
opts.mat_record	record the .mat file?	0 or 1	No or Yes
opts.analyze_mode	analyze the algorithm?	0 or 1	No or Yes
opts.progress	show the progress bar?	0 or 1	No or Yes

Table A.4 *Vector of the analyzed results*

Index	Descriptions
1	Number of periods
2	Strategy's average period return
3	Market's average period return
4	Strategy's winning ratio over the market
5	Alpha (α)
6	Beta (β)
7	*t*-statistics
8	*p*-value
9	Annualized percentage yield
10	Annualized standard deviation
11	Sharpe ratio
12	Drawdown at the end
13	Maximum drawdown during the periods
14	Calmar ratio

A.2.2.2   Examples

Example 1 *Calling a BCRP strategy on the SP500 dataset, mute verbosed outputs:*

```
>> opts.quiet_mode = 1; opts.display_interval = 500;
opts.log_mode = 1; opts.mat_mode = 1;
opts.analyze_mode = 1; opts.progress = 0;
>> manager('bcrp', 'sp500', {0}, opts);
```

Then the algorithm outputs are listed below:

```
>> manager('bcrp', 'sp500', {0}, opts);
----Begin bcrp on sp500-----
------------------------------------
BCRP(tc=0.0000), Final return: 4.07
------------------------------------
----End bcrp on sp500-----
>>
```

Example 2 *Calling a BCRP strategy on the SP500 dataset, display verbosed outputs:*

```
>> opts.quiet_mode = 0; opts.display_interval = 200;
opts.log_mode = 1; opts.mat_mode = 1;
opts.analyze_mode = 1; opts.progress = 0;
>> manager('bcrp', 'sp500', {0}, opts);
```

Then the algorithm outputs are listed below:

```
>> manager('bcrp', 'sp500', {0}, opts);
Running strategy bcrp on dataset sp500
Loading dataset sp500.
Finish loading dataset sp500
The size of the dataset is 1276x25.
Start Time: 2013-0721-13-22-05-664.
----Begin bcrp on sp500-----
-----------------------------------
Parameters [tc:0.000000]
day  Daily Return    Total return
500 1.055339    4.634783
1000    1.018404    4.560191
BCRP(tc=0.0000), Final return: 4.07
-----------------------------------
----End bcrp on sp500-----
Stop Time: 2013-0721-13-22-08-144.
Elapse time(s): 2.486262.
Result Analysis
-----------------------------------
Statistical Test
Size: 1276
MER(Strategy): 0.0015
MER(Market):0.0003
WinRatio:0.5063
Alpha:0.0010
Beta:1.3216
t-statistics:2.1408
p-Value:0.0162
-----------------------------------
Risk Adjusted Return
Volatility Risk analysis
APY: 0.3240
Volatility Risk: 0.4236
Sharpe Ratio: 0.6705
Drawdown analysis
APY: 0.3240
DD: 0.3103
MDD: 0.5066
CR: 0.6395
-----------------------------------
>>
```

A.3   Strategies

This section focuses on describing the implemented strategies in the toolbox. We describe the four implemented categories of algorithms: benchmarks, follow the winner, follow the loser, and pattern matching–based approaches.

A.3.1   Benchmarks

In the financial markets, there exist various benchmarks (such as indices, etc.). In this section, we introduce four benchmarks: Uniform Buy and Hold, Best Stock, Uniform Constant Rebalanced Portfolios, and Best Constant Rebalanced Portfolios.

A.3.1.1   Uniform Buy and Hold

Description   The "buy and hold" (BAH) strategy buys the set of assets at the beginning and holds the allocation of assets till the end of trading periods. BAH with an initial uniform portfolio is termed "uniform buy and hold" (UBAH), which is often a market strategy in the related literature. The final cumulative wealth achieved by a BAH strategy is the initial portfolio weighted average of individual stocks' final wealth,

$$S_n(\text{BAH}(\mathbf{b}_1)) = \mathbf{b}_1 \cdot \left(\bigodot_{t=1}^{n} \mathbf{x}_t \right),$$

where $\mathbf{b}_1$ denotes the initial portfolio. In the case of UBAH, $\mathbf{b}_1 = \left(\frac{1}{m}, \ldots, \frac{1}{m} \right)$. To see its update clearly, BAH's explicit portfolio update can also be written as

$$\mathbf{b}_{t+1} = \frac{\mathbf{b}_t \odot \mathbf{x}_t}{\mathbf{b}_t^\top \mathbf{x}_t}, \tag{A.1}$$

where $\odot$ denotes the operation of element-wise product.

Usage

```
ubah(fid, data, {λ}, opts);
```

- fid: file handle for writing log file;
- data: market sequence matrix;
- $\lambda \in [0, 1)$: proportional transaction cost rate; and
- opts: options for behavioral control.

Example   Call market (uniform BAH) strategy on the "NYSE (O)" dataset with a transaction cost rate of 0.

```
1: >> manager('market', 'nyse-o', {0}, opts);
```

A.3.1.2 Best Stock

Description "Best Stock" (Best) is a special BAH strategy that buys the best stock in hindsight. The final cumulative wealth achieved by the Best strategy can be calculated as

$$S_n(Best) = \max_{\mathbf{b} \in \Delta_m} \mathbf{b} \cdot \left(\bigodot_{t=1}^{n} \mathbf{x}_t \right) = S_n(\text{BAH}(\mathbf{b}^\circ)),$$

where the initial portfolio $\mathbf{b}^\circ$ can be calculated as

$$\mathbf{b}^\circ = \arg \max_{\mathbf{b} \in \Delta_m} \mathbf{b} \cdot \left(\bigodot_{t=1}^{n} \mathbf{x}_t \right).$$

Its portfolio update can also be explicitly written as the same as Equation A.1, except that the initial portfolio equals $\mathbf{b}^\circ$.

Usage

```
best(fid, data, {λ}, opts);
```

- fid: file handle for writing log file;
- data: market sequence matrix;
- $\lambda \in [0, 1)$: transaction costs rate; and
- opts: options for behavioral control.

Example Call Best Stock strategy on the "NYSE (O)" dataset with a transaction cost rate of 0.

```
1: >> manager('best', 'nyse-o', {0}, opts);
```

A.3.1.3 Uniform Constant Rebalanced Portfolios

Description "Constant rebalanced portfolios" (CRP) is a fixed proportion strategy, which rebalances to a preset portfolio at the beginning of every period. In particular, the portfolio strategy can be represented as $\mathbf{b}_1^n = \{\mathbf{b}, \mathbf{b}, \dots\}$. The final cumulative portfolio wealth achieved by a CRP strategy after n periods is defined as

$$S_n(\text{CRP}(\mathbf{b})) = \prod_{t=1}^{n} \mathbf{b}^\top \mathbf{x}_t.$$

In particular, UCRP chooses a uniform portfolio as the preset portfolio, that is, $\mathbf{b} = \left(\frac{1}{m}, \dots, \frac{1}{m} \right)$.

Usage

```
ucrp(fid, data, {λ}, opts);
```

- fid: file handle for writing log file;
- data: market sequence matrix;

- $\lambda \in [0, 1)$: transaction costs rate; and
- opts: options for behavioral control.

Example Call UCRP strategy on the "NYSE (O)" dataset with a transaction cost rate of 0.

```
1: >> manager('ucrp', 'nyse-o', {0}, opts);
```

A.3.1.4 Best Constant Rebalanced Portfolios

Description "Best constant rebalanced portfolio" (BCRP) is a special CRP strategy that sets the portfolio as the portfolio that maximizes the terminal wealth in hindsight. BCRP achieves a final cumulative portfolio wealth as follows:

$$S_n(\text{BCRP}) = \max_{\mathbf{b} \in \Delta_m} S_n(\text{CRP}(\mathbf{b})) = S_n(\text{CRP}(\mathbf{b}^\star)),$$

and its portfolio is calculated in hindsight as

$$\mathbf{b}^\star = \arg\max_{\mathbf{b}^n \in \Delta_m} \log S_n(\text{CRP}(\mathbf{b})) = \arg\max_{\mathbf{b} \in \Delta_m} \sum_{t=1}^{n} \log(\mathbf{b}^\top \mathbf{x}_t).$$

Usage

```
                bcrp(fid, data, {λ}, opts);
```

- fid: file handle for writing log file;
- data: market sequence matrix;
- $\lambda \in [0, 1)$: transaction costs rate; and
- opts: options for behavioral control.

Example Call BCRP strategy on the "NYSE (O)" dataset with a transaction cost rate of 0.

```
1: >> manager('bcrp', 'nyse-o', {0}, opts);
```

A.3.2 Follow the Winner

The Follow the Winner approach is characterized by transferring portfolio weights from the underperforming assets (experts) to the outperforming ones.

A.3.2.1 Universal Portfolios

Description Cover's (1991) "Universal Portfolios" (UP) uniformly buys and holds the whole set of CRP experts within the simplex domain. Its cumulative wealth is calculated as

$$S_n(\text{UP}) = \int_{\Delta_m} S_n(\mathbf{b})d\mu(\mathbf{b}).$$

Moreover, we adopt an implementation (Kalai and Vempala 2002), which is based on nonuniform random walks that are rapidly mixing and which requires a polynomial time.

Usage

```
up(fid, data, {λ}, opts);
```

- fid: file handle for writing log file;
- data: market sequence matrix;
- $\lambda \in [0, 1)$: transaction costs rate; and
- opts: options for behavioral control.

Example   Call Cover's Universal Portfolios on the "NYSE (O)" dataset with default parameters and a transaction cost rate of 0.

```
1: >> manager('up', 'nyse-o', {0}, opts);
```

A.3.2.2   Exponential Gradient

Description   "Exponential gradient" (EG) (Helmbold et al. 1996) tracks the best stock and adopts a regularization term to constrain the deviation from the previous portfolio, that is, EG's formulation is

$$\mathbf{b}_{t+1} = \arg\max_{\mathbf{b} \in \Delta_m} \quad \eta \log \mathbf{b} \cdot \mathbf{x}_t - R(\mathbf{b}, \mathbf{b}_t),$$

where η refers to the learning rate and $R(\mathbf{b}, \mathbf{b}_t)$ denotes relative entropy, or $R(\mathbf{b}, \mathbf{b}_t) = \sum_{i=1}^{m} b_i \log \frac{b_i}{b_{t,i}}$. Solving the optimization, we can obtain EG's portfolio explicit update:

$$b_{t+1,i} = b_{t,i} \exp\left(\eta \frac{x_{t,i}}{\mathbf{b}_t \cdot \mathbf{x}_t}\right)/Z, \quad i = 1, \ldots, m,$$

where Z denotes the normalization term such that the portfolio element sums to 1.

Usage

```
eg(fid, data, {η, λ}, opts);
```

- fid: file handle for writing log file;
- data: market sequence matrix;
- η: learning rate;
- λ: transaction costs rate; and
- opts: options for behavioral control.

Example   Call EG on the "NYSE (O)" dataset with a learning rate of 0.05 and a transaction cost rate of 0.

```
1: >> manager('eg', 'nyse-o', {0.05, 0}, opts);
```

A.3.2.3   Online Newton Step

Description   "Online Newton Step" (ONS) (Agarwal et al. 2006) tracks the best CRP to date and adopts a L2-norm regularization to constrain the portfolio's variability. In particular, its formulation is

$$\mathbf{b}_{t+1} = \arg\max_{\mathbf{b}\in\Delta_m} \sum_{\tau=1}^{t} \log(\mathbf{b}\cdot\mathbf{x}_\tau) - \frac{\beta}{2}\|\mathbf{b}\|.$$

Solving the optimization, we can obtain the explicit portfolio update of ONS:

$$\mathbf{b}_1 = \left(\frac{1}{m}, \ldots, \frac{1}{m}\right), \quad \mathbf{b}_{t+1} = \Pi_{\Delta_m}^{\mathbf{A}_t}(\delta\mathbf{A}_t^{-1}\mathbf{p}_t),$$

with

$$\mathbf{A}_t = \sum_{\tau=1}^{t}\left(\frac{\mathbf{x}_\tau\mathbf{x}_\tau^\top}{(\mathbf{b}_\tau\cdot\mathbf{x}_\tau)^2}\right) + \mathbf{I}_m, \quad \mathbf{p}_t = \left(1+\frac{1}{\beta}\right)\sum_{\tau=1}^{t}\frac{\mathbf{x}_\tau}{\mathbf{b}_\tau\cdot\mathbf{x}_\tau},$$

where β is the trade-off parameter, δ is a scaling term, and $\Pi_{\Delta_m}^{\mathbf{A}_t}(\cdot)$ is an exact projection to the simplex domain.

Usage

```
ons(fid, data, {η, β, δ, λ}, opts)
```

- fid: file handle for writing log file;
- data: market sequence matrix;
- η: mixture parameter;
- β: trade-off parameter;
- δ: heuristic tuning parameter;
- λ: transaction costs rate; and
- opts: options for behavioral control.

Example   Call the ONS on the "NYSE (O)" dataset with a transaction cost rate of 0.

```
1: >> manager('ons', 'nyse-o', {0, 1, 1/8, 0}, opts);
```

A.3.3   Follow the Loser

The Follow the Loser approaches assume that the underperforming assets will revert and outperform others in the subsequent periods. Thus, their common behavior is to move portfolio weights from the outperforming assets to the underperforming assets.

A.3.3.1   Anticorrelation

Description   "Anticorrelation" (Anticor) (Borodin et al. 2004) transfers the wealth from the outperforming stocks to the underperforming stocks via their cross-correlation and autocorrelation. Anticor adopts logarithmic price relatives in two specific market windows, that is, $\mathbf{y}_1 = \log(\mathbf{x}_{t-2w+1}^{t-w})$ and $\mathbf{y}_2 = \log(\mathbf{x}_{t-w+1}^{t})$. It then calculates the cross-correlation matrix between $\mathbf{y}_1$ and $\mathbf{y}_2$:

$$M_{cov}(i, j) = \frac{1}{w-1}(\mathbf{y}_{1,i} - \bar{y}_1)^\top (\mathbf{y}_{2,j} - \bar{y}_2),$$

$$M_{cor}(i, j) = \begin{cases} \frac{M_{cov}(i,j)}{\sigma_1(i)*\sigma_2(j)} & \sigma_1(i), \sigma_2(j) \neq 0 \\ 0 & \text{otherwise} \end{cases}.$$

Then, following the cross-correlation matrix, Anticor moves the proportions from the stocks increased more to the stocks increased less, in which the corresponding amounts are adjusted according to the cross-correlation matrix. In particular, if asset i increases more than asset j and their sequences in the window are positively correlated, Anticor claims a transfer from asset i to j with the amount equaling the crosscorrelation value ($M_{cor}(i, j)$) minus their negative autocorrelation values ($\min\{0, M_{cor}(i, i)\}$ and $\min\{0, M_{cor}(j, j)\}$). These transfer claims are finally normalized to keep the portfolio in the simplex domain.

Usage We implemented two Anticor algorithms, BAH$_W$(Anticor) and BAH$_W$ (Anticor(Anticor)). Their usages are listed below.

```
anticor(fid, data, {W, λ}, opts);
anticor_anticor(fid, data, {W, λ}, opts);
```

- fid: file handle for writing log file;
- data: market sequence matrix;
- W: maximal window size;
- λ: transaction cost rates; and
- opts: options for behavioral control.

Example Call both Anticor algorithms on the "NYSE (O)" dataset with a maximal window size of 30 and a transaction cost rate of 0.

```
1: >> manager('anticor', 'nyse-o', {30, 0}, opts);
2: >> manager('anticor_anticor', 'nyse-o', {30, 0}, opts);
```

A.3.3.2 Passive–Aggressive Mean Reversion

Description Rather than tracking the best stock, "passive–aggressive mean reversion" (PAMR) (Li et al. 2012) explicitly tracks the worst stocks, while adopting regularization techniques to constrain the deviation from the last portfolio. In particular, PAMR's formulation is

$$\mathbf{b}_{t+1} = \underset{\mathbf{b} \in \Delta_m}{\arg\min} \frac{1}{2} \|\mathbf{b} - \mathbf{b}_t\|^2 \quad \text{s.t.} \quad \ell_\epsilon(\mathbf{b}; \mathbf{x}_t) = 0,$$

where $\ell_\epsilon(\mathbf{b}; \mathbf{x}_t)$ denotes a predefined loss function to capture the mean reversion property,

$$\ell_\epsilon(\mathbf{b}; \mathbf{x}_t) = \begin{cases} 0 & \mathbf{b} \cdot \mathbf{x}_t \leq \epsilon \\ \mathbf{b} \cdot \mathbf{x}_t - \epsilon & \text{otherwise} \end{cases}.$$

Solving the optimization, we can obtain PAMR's portfolio update:

$$\mathbf{b}_{t+1} = \mathbf{b}_t - \tau_t(\mathbf{x}_t - \bar{x}_t \mathbf{1}), \quad \tau_t = \max\left\{0, \frac{\mathbf{b}_t \cdot \mathbf{x}_t - \epsilon}{\|\mathbf{x}_t - \bar{x}_t \mathbf{1}\|^2}\right\}.$$

Usage   We implemented three PAMR algorithms (i.e., PAMR, PAMR-I, and PAMR-II). Their usages are listed below.

```
pamr(fid, data, {ε, λ}, opts);
pamr_1(fid, data, {ε, C, λ}, opts);
pamr_2(fid, data, {ε, C, λ}, opts);
```

- fid: file handle for writing log file;
- data: market sequence matrix;
- ϵ: mean reversion threshold;
- C: aggressive parameter;
- λ: transaction cost rates; and
- opts: options for behavioral control.

Example   Call the three PAMR algorithms on the "NYSE (O)" dataset with a mean reversion threshold of 0.5, an aggressive parameter of 30, and a transaction cost rate of 0.

```
1: >> manager('pamr', 'nyse-o', {0.5, 0}, opts);
2: >> manager('pamr_1', 'nyse-o', {0.5, 500, 0}, opts);
3: >> manager('pamr_2', 'nyse-o', {0.5, 500, 0}, opts);
```

A.3.3.3   Confidence-Weighted Mean Reversion

Description   "Confidence-weighted mean reversion" (CWMR) (Li et al. 2013) models the portfolio vector on a Gaussian distribution and explicitly updates the distribution following the mean reversion principle. In particular, CWMR's formulation is

$$(\mu_{t+1}, \Sigma_{t+1}) = \underset{\mu \in \Delta_m, \Sigma}{\arg\min} \quad D_{KL}(\mathcal{N}(\mu, \Sigma) \| \mathcal{N}(\mu_t, \Sigma_t))$$

$$\text{s.t.} \quad \Pr[\mu \cdot \mathbf{x}_t \leq \epsilon] \geq \theta.$$

Expanding the constraint, the resulting optimization problem is not convex. The authors provided two methods to solve the optimization (i.e., CWMR-Var and CWMR-Stdev). CWMR-Var involves linearizing the constraint and solving the resulting optimization, and one can obtain the closed form update scheme as

$$\mu_{t+1} = \mu_t - \lambda_{t+1} \Sigma_t (\mathbf{x}_t - \bar{x}_t \mathbf{1}), \quad \Sigma_{t+1}^{-1} = \Sigma_t^{-1} + 2\lambda_{t+1} \phi \mathbf{x}_t \mathbf{x}_t^\top,$$

where λ_{t+1} corresponds to the Lagrangian multiplier calculated by Eq. (11) in Li et al. (2013), and $\bar{x}_t = \frac{\mathbf{1}^\top \Sigma_t \mathbf{x}_t}{\mathbf{1}^\top \Sigma_t \mathbf{1}}$ denotes the confidence-weighted price relative average. CWMR-Stdev involves the decomposition of the covariance matrix and can also release similar portfolio update formulas.

Usage   We implemented two CWMR algorithms (i.e., CWMR-Var and CWMR-Stdev). Their usages are listed below.

```
cwmr_var(fid, data, {ϕ, ε, λ}, opts);
cwmr_stdev(fid, data, {ϕ, ε, λ}, opts);
```

- fid: file handle for writing log file;
- data: market sequence matrix;
- ϕ: confidence parameter;
- ϵ: mean reversion threshold;
- λ: transaction cost rates; and
- opts: options for behavioral control.

Example   Call the two CWMR algorithms on the "NYSE (O)" dataset with a confidence parameter of 2, a mean reversion parameter of 0.5, and a transaction cost rate of 0.

```
1: >> manager('cwmr_var', 'nyse-o', {2, 0.5, 0}, opts);
2: >> manager('cwmr_stdev', 'nyse-o', {2, 0.5, 0}, opts);
```

A.3.3.4   Online Moving Average Reversion

Description   "Online moving average reversion" (OLMAR) (Li and Hoi 2012) explicitly predicts next price relatives following the mean reversion idea (i.e., MAR-1 borrows the simple moving average):

$$\tilde{\mathbf{x}}_{t+1}(w) = \frac{1}{w}\left(1 + \frac{1}{\mathbf{x}_t} + \cdots + \frac{1}{\bigodot_{i=0}^{w-2} \mathbf{x}_{t-i}}\right),$$

where w is the window size and $\odot$ denotes the element-wise product; and MAR-2 borrows the exponential moving average:

$$\tilde{\mathbf{x}}_{t+1}(\alpha) = \alpha\mathbf{1} + (1-\alpha)\frac{\tilde{\mathbf{x}}_t}{\mathbf{x}_t},$$

where $\alpha \in (0, 1)$ denotes the decaying factor and the operations are all element-wise. Then, OLMAR's formulation is

$$\mathbf{b}_{t+1} = \underset{\mathbf{b} \in \Delta_m}{\arg\min} \quad \frac{1}{2}\|\mathbf{b} - \mathbf{b}_t\|^2 \text{ s.t. } \mathbf{b} \cdot \tilde{\mathbf{x}}_{t+1} \geq \epsilon.$$

Solving the optimization, we can obtain its portfolio update:

$$\mathbf{b}_{t+1} = \mathbf{b}_t + \lambda_{t+1}(\tilde{\mathbf{x}}_{t+1} - \bar{x}_{t+1}\mathbf{1}),$$

where $\bar{x}_{t+1} = \frac{1}{m}(\mathbf{1} \cdot \tilde{\mathbf{x}}_{t+1})$ denotes the average predicted price relative and λ_{t+1} is the Lagrangian multiplier calculated as

$$\lambda_{t+1} = \max\left\{0, \frac{\epsilon - \mathbf{b}_t \cdot \tilde{\mathbf{x}}_{t+1}}{\|\tilde{\mathbf{x}}_{t+1} - \bar{x}_{t+1}\mathbf{1}\|^2}\right\}.$$

Usage   We implemented two OLMAR algorithms (i.e., OLMAR-I and OLMAR-II). Their usages are listed below.

```
olmar1(fid, data, {ε, W, λ}, opts);
olmar2(fid, data, {ε, α, λ}, opts);
```

- fid: file handle for writing log file;
- data: market sequence matrix;
- ϵ: mean reversion threshold;
- W: window size for simple moving average;
- $\alpha \in [0, 1]$: decaying factor to calculate exponential moving average;
- $\lambda \in [0, 1)$: transaction cost rates; and
- opts: options for behavioral control.

Example   Call the two OLMAR algorithms on the "NYSE (O)" dataset with a mean reversion threshold of 10, a window size of 5, a decaying factor of 0.5, and a transaction cost rate of 0.

```
1: >> manager('olmar1', 'nyse-o', {10, 5, 0}, opts);
2: >> manager('olmar2', 'nyse-o', {10, 0.5, 0}, opts);
```

A.3.4 Pattern Matching–Based Approaches

The pattern matching–based approaches are based on the assumption that market sequences with similar preceding market appearances tend to reappear. Thus, the common behavior of these approaches is to first identify similar market sequences that are deemed similar to the coming sequence, and then obtain a portfolio that maximizes the expected return based on these similar sequences. Algorithm A.3 illustrates the first step, or the sample selection procedure. The second step, or the portfolio optimization procedure, often follows the following optimization:

$$\mathbf{b}_{t+1} = \arg\max_{\mathbf{b} \in \Delta_m} \prod_{i \in C(\mathbf{x}_1^t)} \mathbf{b} \cdot \mathbf{x}_i. \qquad (A.2)$$

A.3.4.1 Nonparametric Kernel-Based Log-Optimal Strategy

Description "Nonparametric *kernel-based* sample selection" (B^K) (Györfi et al. 2006) identifies the similarity set by comparing two market windows via Euclidean distance:

$$C_K(\mathbf{x}_1^t, w) = \left\{ w < i < t+1 : \|\mathbf{x}_{t-w+1}^t - \mathbf{x}_{i-w}^{i-1}\| \leq \frac{c}{\ell} \right\},$$

Algorithm A.3: Sample selection framework ($C(\mathbf{x}_1^t, w)$).

Input: $\mathbf{x}_1^t$: Historical market sequence; w: window size;
Output: C: Index set of similar price relatives.

Initialize $C = \emptyset$;
if $t \leq w+1$ **then**
| return;
end
for $i = w+1, w+2, \ldots, t$ **do**
| **if** $\mathbf{x}_{i-w}^{i-1}$ *is similar to* $\mathbf{x}_{t-w+1}^t$ **then**
| | $C = C \cup \{i\}$;
| **end**
end

---

where c and ℓ are the thresholds used to control the number of similar samples. Then, it obtains an optimal portfolio via solving Equation A.2.

Usage

```
bk_run(fid, data, {K, L, c, λ}, opts);
```

- fid: file handle for writing log file;
- data: market sequence matrix;
- K: maximal window size;
- L: used to split the parameter space of each k;

- c: similarity threshold;
- $\lambda \in [0, 1)$: transaction cost rates; and
- opts: options for behavioral control.

Example   Call the B^K algorithm on the "NYSE (O)" dataset with default parameters and a transaction cost rate of 0.

```
1: >> manager('bk', 'nyse-o', {5, 10, 1, 0}, opts);
```

A.3.4.2   Nonparametric Nearest-Neighbor Log-Optimal Strategy

Description   "Nonparametric *nearest-neighbor-based* sample selection" (B^{NN}) (Györfi et al. 2008) searches the price relatives whose preceding market windows are within the ℓ nearest neighbor of latest market window in terms of Euclidean distance:

$$C_N(\mathbf{x}_1^t, w) = \{w < i < t+1 : \mathbf{x}_{i-w}^{i-1} \text{ is among the } \ell \text{ NNs of } \mathbf{x}_{t-w+1}^t\},$$

where ℓ is a threshold parameter. Then, the strategy obtains an optimal portfolio via solving Equation A.2.

Usage

$$\text{bnn(fid, data, \{K, L, } \lambda\text{\}, opts)}$$

- fid: file handle for writing log file;
- data: market sequence matrix;
- K: maximal window size;
- L: parameter to split the parameter space of each k;
- $\lambda \in [0, 1)$: transaction cost rates; and
- opts: options for behavioral control.

Example   Call the B^{NN} algorithm on the "NYSE (O)" dataset with default parameters and a transaction cost rate of 0.

```
1: >> manager('bnn', 'nyse-o', {5, 10, 0}, opts);
```

A.3.4.3   Correlation-Driven Nonparametric Learning Strategy

"*Correlation-driven* nonparametric sample selection" (CORN) (Li et al. 2011a) identifies the similarity among two market windows via a correlation coefficient:

$$C_C(\mathbf{x}_1^t, w) = \left\{ w < i < t+1 : \frac{cov(\mathbf{x}_{i-w}^{i-1}, \mathbf{x}_{t-w+1}^t)}{std(\mathbf{x}_{i-w}^{i-1})std(\mathbf{x}_{t-w+1}^t)} \geq \rho \right\},$$

where ρ is a predefined threshold. Then, it obtains an optimal portfolio via solving Equation A.2.

Usage

```
corn(fid, data, {w, c, λ}, opts);
cornu(fid, data, {K, L, c, λ}, opts);
cornk_run(fid, data, {K, L, pc, λ}, opts)
```

- fid: file handle for writing log file;
- data: market sequence matrix;
- w: window size;
- K: maximal window size;
- L: used to split the parameter space of each k;
- c: correlation threshold;
- pc: percentage of experts to be selected;
- λ ∈ [0, 1): transaction cost rates; and
- opts: options for behavioral control.

Example Below we call three CORN algorithms with their default parameters.

```
1: >> manager('corn', 'nyse-o', {5, 0.1, 0}, opts);
2: >> manager('cornu', 'nyse-o', {5, 1, 0.1, 0}, opts);
3: >> manager('cornk', 'nyse-o', {5, 10, 0.1, 0}, opts);
```

A.4 Summary

In this manual, we describe the OLPS toolbox in detail. OLPS is the first toolbox for the research of OLPS problems. It is easy to use and can be extended to include new algorithms and datasets. We hope this toolbox can facilitate further research on this topic.

Appendix B

Proofs and Derivations

B.1 Proof of CORN

B.1.1 Proof of Theorem 1*

In this appendix, we give a detailed proof that the portfolio scheme correlation-driven nonparametric learning (CORN) (Li et al. 2011a) is universal with respect to the class of all ergodic processes. We first give a concise definition about "universal" considered in this note.

Definition 1 *An investment strategy* **B** *is called universal with respect to a class of stationary and ergodic processes* $\{\mathbf{X}_n\}_{-\infty}^{+\infty}$, *if, for each process in the class,*

$$\lim_{n \to \infty} \frac{1}{n} \log S_n(\mathbf{B}) = W^* \quad \text{almost surely.}$$

Before we give the theorem and its proof, we introduce some necessary lemmas.

Lemma B.1 (Breiman 1957 [Correction version 1960]) Let $Z = \{Z_i\}_{-\infty}^{\infty}$ be a stationary and ergodic process. For each positive integer i, let T^i denote the operator that shifts any sequence $\{\ldots, z_{-1}, z_0, z_1, \ldots\}$ by i digits to the left. Let $f_1, f_2, \ldots$ be a sequence of real-valued functions such that $\lim_{n \to \infty} f_n(Z) = f(Z)$ almost surely for some function f. Assume that $\mathbb{E} \sup_n |f_n(Z)| < \infty$. Then,

$$\lim_{n \to \infty} \frac{1}{n} \sum_{i=1}^{n} f_i(T^i Z) = \mathbb{E} f(Z) \quad \text{almost surely.}$$

Lemma B.2 (Algoet and Cover 1988) Let $\mathbf{Q}_{n \in \mathcal{N} \cup \{\infty\}}$ be a family of regular probability distributions over the set $\mathbb{R}_+^d$ of all market vectors such that $E\{|\log U_n^{(j)}|\} < \infty$ for any coordinate of a random market vector $\mathbf{U}_n = (U_n^{(1)}, \ldots, U_n^{(d)})$ distributed according to $\mathbf{Q}_n$. In addition, let $\mathbf{B}^*(\mathbf{Q}_n)$ be the set of all log-optimal portfolios with

---

*The proof idea is mainly provided by Vladimir Vovk, and the proof is then finished by Dingjiang Huang and Bin Li.

respect to $\mathbf{Q}_n$, that is, the set of all portfolios $\mathbf{b}$ that attain $\max_{\mathbf{b} \in \Delta_d} E\{\log\langle \mathbf{b}, \mathbf{U}_n\rangle\}$. Consider an arbitrary sequence $\mathbf{b}_n \in \mathbf{B}^*(\mathbf{Q}_n)$. If

$$\mathbf{Q}_n \to \mathbf{Q}_\infty \quad \text{weakly as } n \to \infty,$$

then, for $\mathbf{Q}_\infty$-almost all $\mathbf{u}$,

$$\lim_{n \to \infty} \langle \mathbf{b}_n, \mathbf{u}\rangle \to \langle \mathbf{b}^*, \mathbf{u}\rangle,$$

where the right-hand side is constant as $\mathbf{b}^*$ ranges over $B^*(\mathbf{Q}_\infty)$.

Lemma B.3 (Algoet and Cover 1988) Let $\mathbf{X}$ be a random market vector defined on a probability space$(\Omega, \mathcal{F}, \mathbb{P})$ satisfying $E\{|\log X^{(j)}|\} < \infty$. If $\mathcal{F}_k$ is an increasing sequence of sub-σ-fields of $\mathcal{F}$ with:

$$\mathcal{F}_k \nearrow \mathcal{F}_\infty \subseteq \mathcal{F},$$

then

$$\mathbb{E}\left\{ \max_{\mathbf{b}} \mathbb{E}[\log\langle \mathbf{b}, \mathbf{X}\rangle | \mathcal{F}_k] \right\} \nearrow \mathbb{E}\left\{ \max_{\mathbf{b}} \mathbb{E}[\log\langle \mathbf{b}, \mathbf{X}\rangle | \mathcal{F}_\infty] \right\},$$

as $k \to \infty$ where the maximum on the left-hand side is taken on over all $\mathcal{F}_k$-measurable functions $\mathbf{b}$, and the maximum on the right-hand side is taken on over all $\mathcal{F}_\infty$-measurable functions $\mathbf{b}$.

Lemma B.4 Let μ be the Lebesgue measure on the Euclidean space $\mathbf{R}^n$ and A be a Lebesgue measurable subset of $\mathbf{R}^n$. Define the approximate density of A in a ε-neighborhood of a point x in $\mathbf{R}^n$ as

$$d_\varepsilon(x) = \frac{\mu(A \cap B_\varepsilon(x))}{\mu(B_\varepsilon(x))},$$

where B_ε denotes the closed ball of radius ε centered at x. Then for almost every point x of A the density,

$$d(x) = \lim_{\varepsilon \to 0} d_\varepsilon(x)$$

exists and is equal to 1.

Lemma B.5 The inequality

$$\frac{\text{cov}(\mathbf{X}, \mathbf{X}')}{\sqrt{\text{Var}(\mathbf{X})}\sqrt{\text{Var}(\mathbf{X}')}} \geq \rho,$$

which describes the similarity of $\mathbf{X}$ and $\mathbf{X}'$ in CORN strategy, is approximately equivalent to

$$2\text{Var}(\mathbf{X})(1 - \rho) \geq \mathbb{E}\{(\mathbf{X} - \mathbf{X}')^2\}.$$

Proof *In general, from the covariance* $\text{cov}(\mathbf{X}, \mathbf{X}')$ *it is impossible to derive a topology, since* $\text{cov}(\mathbf{X}, \mathbf{X}') = 1$ *does not imply that* $\mathbb{E}\{(\mathbf{X} - \mathbf{X}')^2\} = 0$. *However, because* $\mathbf{X}$ *and* $\mathbf{X}'$ *are relative prices, then we have* $\mathbb{E}\{(\mathbf{X} - \mathbf{X}')^2\} \approx 0$. *For the Euclidean distance, we have that*

$$\mathbb{E}\{(\mathbf{X} - \mathbf{X}')^2\} = \text{Var}(\mathbf{X} - \mathbf{X}') + (\mathbb{E}\{\mathbf{X} - \mathbf{X}'\})^2$$
$$= \text{Var}(\mathbf{X}) - 2\text{cov}(\mathbf{X}, \mathbf{X}') + \text{Var}(\mathbf{X}') + (\mathbb{E}\{\mathbf{X} - \mathbf{X}'\})^2.$$

Thus, the similarity means that

$$\frac{\text{Var}(\mathbf{X}) + \text{Var}(\mathbf{X}') + (\mathbb{E}\{\mathbf{X} - \mathbf{X}'\})^2 - \mathbb{E}\{(\mathbf{X} - \mathbf{X}')^2\}}{\sqrt{\text{Var}(\mathbf{X})}\sqrt{\text{Var}(\mathbf{X}')}} \geq 2\rho$$

or, equivalently,

$$\text{Var}(\mathbf{X}) + \text{Var}(\mathbf{X}') + (\mathbb{E}\{\mathbf{X} - \mathbf{X}'\})^2 - 2\rho\sqrt{\text{Var}(\mathbf{X})}\sqrt{\text{Var}(\mathbf{X}')} \geq \mathbb{E}\{(\mathbf{X} - \mathbf{X}')^2\}.$$

Since both $\text{Var}(\mathbf{X})$ *and* $|\mathbb{E}\{\mathbf{X} - \mathbf{X}'\}|$ *have the same order of magnitude,* *they are in the range* $10^{-4}, 10^{-3}$; *therefore, the previous inequality approximately means that*

$$2\text{Var}(\mathbf{X})(1 - \rho) \geq \mathbb{E}\{(\mathbf{X} - \mathbf{X}')^2\}.$$

Lemma B.6 Assume that $\mathbf{x}_1, \mathbf{x}_2, \ldots$ are the realizations of the random vectors $\mathbf{X}_1, \mathbf{X}_2, \ldots$ drawn from the vector-valued stationary and ergodic process $\{\mathbf{X}_n\}_{-\infty}^{\infty}$. The fundamental limits (determined in Algoet 1992, 1994; Algoet and Cover 1988), reveal that the so-called *log-optimum portfolio* $\mathbf{B}^* = \{\mathbf{b}^*(\cdot)\}$ is the best possible choice. More precisely, in trading period n, let $\mathbf{b}^*(\cdot)$ be such that

$$\mathbb{E}\{\log\langle \mathbf{b}^*(\mathbf{X}_1^{n-1}), \mathbf{X}_n\rangle | \mathbf{X}_1^{n-1}\} = \max_{\mathbf{b}(\cdot)} \mathbb{E}\{\log\langle b(\mathbf{X}_1^{n-1}), \mathbf{X}_n\rangle | \mathbf{X}_1^{n-1}\}.$$

If $S_n^* = S_n(\mathbf{B}^*)$ denotes the capital achieved by a log-optimum portfolio strategy $\mathbf{B}^*$, after n trading periods, then for any other investment strategy $\mathbf{B}$ with capital $S_n = S_n(\mathbf{B})$ and for any stationary and ergodic process $\{\mathbf{X}_n\}_{-\infty}^{\infty}$,

$$\limsup_{n \to \infty} \frac{1}{n} \log \frac{S_n}{S_n^*} \leq 0 \quad \text{almost surely}$$

and

$$\lim_{n \to \infty} \frac{1}{n} \log S_n^* = W^* \quad \text{almost surely},$$

where

$$W^* = \mathbb{E}\left\{ \max_{\mathbf{b}(\cdot)} \mathbb{E}\{\log\langle \mathbf{b}(\mathbf{X}_{-\infty}^{-1}), \mathbf{X}_0\rangle | \mathbf{X}_{-\infty}^{-1}\}\right\},$$

is the maximal possible growth rate of any investment strategy.

---

*See Appendix C for more details.

Now, we give the universal theorem and its proof.

Theorem B.1 The portfolio scheme CORN is universal with respect to the class of all ergodic processes such that $\mathbb{E}\{|\log X^{(j)}|\} < \infty$, for $j = 1, 2, \ldots, d$.

Proof *To prove that the strategy CORN is universal with respect to the class of all ergodic processes, we need to prove that if, for each process in the class,*

$$\lim_{n \to \infty} \frac{1}{n} \log S_n(\mathbf{B}) = W^* \quad \text{almost surely,}$$

where $\mathbf{B}$ denote the strategy CORN; and

$$W^* = \lim_{n \to \infty} \frac{1}{n} \log S_n^* = \mathbb{E}\{\max_{\mathbf{b}(\cdot)} \mathbb{E}\{\log\langle \mathbf{b}(\mathbf{X}_{-\infty}^{-1}), \mathbf{X}_0\rangle | \mathbf{X}_{-\infty}^{-1}\}\}.$$

We divide the proof into three parts.

(i) According to Lemma B.6, we know that $\lim_{n \to \infty} \left(\frac{1}{n} \log S_n - \frac{1}{n} \log S_n^*\right) \leq 0$, then $\lim_{n \to \infty} \frac{1}{n} \log S_n \leq \lim_{n \to \infty} \frac{1}{n} \log S_n^* = W^*$. So it suffices to prove that

$$\liminf_{n \to \infty} W_n(\mathbf{B}) = \liminf_{n \to \infty} \frac{1}{n} \log S_n(\mathbf{B}) \geq W^* \quad \text{almost surely.}$$

Without loss of generality, we may assume $S_0 = 1$, so that

$$\begin{aligned}
W_n(\mathbf{B}) &= \frac{1}{n} \log S_n(\mathbf{B}) \\
&= \frac{1}{n} \log \left(\sum_{\omega,\rho} q_{\omega,\rho} S_n(\epsilon^{(\omega,\rho)})\right) \\
&\geq \frac{1}{n} \log \left(\sup_{\omega,\rho} q_{\omega,\rho} S_n(\epsilon^{(\omega,\rho)})\right) \\
&= \frac{1}{n} \sup_{\omega,\rho} (\log q_{\omega,\rho} + \log S_n(\epsilon^{(\omega,\rho)})) \\
&= \sup_{\omega,\rho} \left(W_n(\epsilon^{(\omega,\rho)}) + \frac{\log q_{\omega,\rho}}{n}\right).
\end{aligned}$$

Thus,

$$\begin{aligned}
\liminf_{n \to \infty} W_n(\mathbf{B}) &= \liminf_{n \to \infty} \sup_{\omega,\rho} \left(W_n(\epsilon^{(\omega,\rho)}) + \frac{\log q_{\omega,\rho}}{n}\right) \\
&\geq \sup_{\omega,\rho} \liminf_{n \to \infty} \left(W_n(\epsilon^{(\omega,\rho)}) + \frac{\log q_{\omega,\rho}}{n}\right) \\
&= \sup_{\omega,\rho} \liminf_{n \to \infty} W_n(\epsilon^{(\omega,\rho)}). \qquad (B.1)
\end{aligned}$$

The simple argument above shows that the asymptotic rate of growth of the strategy $\mathbf{B}$ is at least as large as the supremum of the rates of growth of all elementary strategies $\epsilon^{(\omega,\rho)}$. Thus, to estimate $\liminf_{n \to \infty} W_n(\mathbf{B})$, it suffices to investigate *the performance of expert $\epsilon^{(\omega,\rho)}$ on the stationary and ergodic market sequence* $\mathbf{X}_0, \mathbf{X}_{-1}, \mathbf{X}_{-2}, \ldots$.

(ii) First, let the integers ω, ρ, and the vector $\mathbf{s} = \mathbf{s}_{-\omega}^{-1} \in \mathbb{R}_+^{d\omega}$ be fixed. From Lemma B.5, we can get that the set $\{X_i : 1 - j + \omega \leq i \leq 0, \dfrac{\text{cov}(\mathbf{X}_{i-\omega}^{i-1},\mathbf{s})}{\sqrt{\text{Var}(\mathbf{X}_{i-\omega}^{i-1})}\sqrt{Var(\mathbf{s})}} \geq \rho\}

can be expressed as $\{X_i : 1 - j + \omega \leq i \leq 0, \mathbb{E}\{(\mathbf{X}_{i-\omega}^{i-1} - \mathbf{s})^2\} \leq 2\text{Var}(\mathbf{s})(1 - \rho)\}$.

Let $\mathbb{P}_{j,\mathbf{s}}^{(\omega,\rho)}$ denote the (random) measure concentrated on $\{X_i : 1 - j + \omega \leq i \leq 0, \mathbb{E}\{(\mathbf{X}_{i-\omega}^{i-1} - \mathbf{s})^2\} \leq 2\text{Var}(\mathbf{s})(1 - \rho)$, defined by

$$\mathbb{P}_{j,\mathbf{s}}^{(\omega,\rho)}(A) = \frac{\displaystyle\sum_{i:1-j+\omega\leq i\leq 0, \mathbb{E}\{(\mathbf{X}_{i-\omega}^{i-1}-\mathbf{s})^2\}\leq 2\text{Var}(\mathbf{s})(1-\rho)} II_A(\mathbf{X}_i)}{|\{i : 1 - j + \omega \leq i \leq 0, \mathbb{E}\{(\mathbf{X}_{i-\omega}^{i-1} - \mathbf{s})^2\} \leq 2\text{Var}(\mathbf{s})(1 - \rho)\}|}, \quad A \subset \mathbb{R}_+^d$$

where II_A denotes the indicator of function of the set A. If the above set of $\mathbf{X}_i's$ is empty, then let $\mathbb{P}_{j,\mathbf{s}}^{(\omega,\rho)} = \delta_{(1,\ldots,1)}$ be the probability measure concentrated on the vector $(1, \ldots, 1)$. In other words, $\mathbb{P}_{j,\mathbf{s}}^{(\omega,\rho)}(A)$ is the relative frequency of the vectors among $\mathbf{X}_{1-j+\omega}, \ldots, \mathbf{X}_0$ that fall in the set A.

Observe that for all $\mathbf{s}$, without probability 1,

$$\mathbb{P}_{j,\mathbf{s}}^{(\omega,\rho)} \to \mathbb{P}_{\mathbf{s}}^{*(\omega,\rho)}$$

$$= \begin{cases} \mathbb{P}_{\mathbf{X}_0 | \mathbb{E}\{(\mathbf{X}_{-\omega}^{-1}-\mathbf{s})^2\}\leq 2\text{Var}(\mathbf{s})(1-\rho)} & \text{if } \mathbb{P}(\mathbb{E}\{(\mathbf{X}_{-\omega}^{-1} - \mathbf{s})^2\} \leq 2\text{Var}(\mathbf{s})(1 - \rho)) > 0 \\ \delta_{(1,\ldots,1)} & \text{if } \mathbb{P}(\mathbb{E}\{(\mathbf{X}_{-\omega}^{-1} - \mathbf{s})^2\} \leq 2\text{Var}(\mathbf{s})(1 - \rho)) = 0 \end{cases}$$

$$\text{(B.2)}$$

weakly as $j \to \infty$, where $\mathbb{P}_{\mathbf{s}}^{*(\omega,\rho)}$ denotes the limit distribution of $\mathbb{P}_{j,\mathbf{s}}^{(\omega,\rho)}$, and $\mathbb{P}_{\mathbf{X}_0 | \mathbb{E}\{(\mathbf{X}_{-\omega}^{-1}-\mathbf{s})^2\}\leq 2\text{Var}(\mathbf{s})(1-\rho)}$ denotes the distribution of the vector $\mathbf{X}_0$ conditioned on the event $\mathbb{E}\{(\mathbf{X}_{-\omega}^{-1} - \mathbf{s})^2\} \leq 2\text{Var}(\mathbf{s})(1 - \rho)$. To see this, let f be a bounded continuous function defined on $\mathbb{R}_+^d$. Then, the ergodic theorem implies that if $\mathbb{P}(\mathbb{E}\{(\mathbf{X}_{-\omega}^{-1} - \mathbf{s})^2\} \leq 2\text{Var}(\mathbf{s})(1 - \rho)) > 0$, then

$$\int f(\mathbf{x})\mathbb{P}_{j,\mathbf{s}}^{*(\omega,\rho)}(d\mathbf{x}) = \frac{\frac{1}{|1-j+\omega|}\displaystyle\sum_{i:1-j+\omega\leq i\leq 0, \mathbb{E}\{(\mathbf{X}_{i-\omega}^{i-1}-\mathbf{s})^2\}\leq 2\text{Var}(\mathbf{s})(1-\rho)} f(\mathbf{X}_i)}{\frac{1}{|1-j+\omega|}|\{i:1-j+\omega\leq i\leq 0, \mathbb{E}\{(\mathbf{X}_{i-\omega}^{i-1}-\mathbf{s})^2\}\leq 2\text{Var}(\mathbf{s})(1-\rho)\}|}$$

$$\to \frac{\mathbb{E}\{f(X_0)II_{\{\mathbb{E}\{(\mathbf{X}_{-\omega}^{-1}-\mathbf{s})^2\}\leq 2\text{Var}(\mathbf{s})(1-\rho)\}}\}}{\mathbb{P}\{\mathbb{E}\{(\mathbf{X}_{-\omega}^{-1}-\mathbf{s})^2\}\leq 2\text{Var}(\mathbf{s})(1-\rho)\}}$$

$$= E\{f(X_0)|\mathbb{E}\{(\mathbf{X}_{-\omega}^{-1} - \mathbf{s})^2\} \leq 2\text{Var}(\mathbf{s})(1 - \rho)\}$$

$$= \int f(\mathbf{x})\mathbb{P}_{\mathbf{X}_0 | \mathbb{E}\{(\mathbf{X}_{-\omega}^{-1}-\mathbf{s})^2\}\leq 2\text{Var}(\mathbf{s})(1-\rho)} \quad \text{almost surely, as } j \to \infty.$$

On the other hand, if $\mathbb{P}(\mathbb{E}\{(\mathbf{X}_{-\omega}^{-1} - \mathbf{s})^2\} \leq 2\text{Var}(\mathbf{s})(1 - \rho)) = 0$, then with probability 1, $\mathbb{P}_{j,\mathbf{s}}^{(\omega,\rho)}$ is concentrated on $(1, \ldots, 1)$ for all j, and $\int f(\mathbf{x})\mathbb{P}_{j,\mathbf{s}}^{(\omega,\rho)}(d\mathbf{x}) = f(1, \ldots, 1)$.

Recall that by definition, $\mathbf{b}^{(\omega,\rho)}(\mathbf{X}_{1-j}^{-1}, \mathbf{s})$ *is a log-optimal portfolio with respect to the probability measure* $\mathbb{P}_{j,s}^{(\omega,\rho)}$. *Let* $\mathbf{b}_{\omega,\rho}^{*}(\mathbf{s})$ *denote a log-optimal portfolio with respect to the limit distribution* $\mathbb{P}_{s}^{*(\omega,\rho)}$. *Then, using Lemma B.2, we infer from Equation B.2 that, as j tends to infinity, we have the almost sure convergence*

$$\lim_{j\to\infty} \langle \mathbf{b}^{(\omega,\rho)}(\mathbf{X}_{1-j}^{-1}, \mathbf{s}), \mathbf{x}_0 \rangle = \langle \mathbf{b}_{\omega,\rho}^{*}(\mathbf{s}), \mathbf{x}_0 \rangle,$$

for $\mathbb{P}_{s}^{*(\omega,\rho)}$ *(almost all* $\mathbf{x}_0$*) and hence for* $\mathbb{P}_{\mathbf{X}_0}$ *(almost all* $\mathbf{x}_0$*). Since* $\mathbf{s}$ *was arbitrary, we obtain*

$$\lim_{j\to\infty} \langle \mathbf{b}^{(\omega,\rho)}(\mathbf{X}_{1-j}^{-1}, \mathbf{X}_{-\&}^{-1}), \mathbf{x}_0 \rangle = \langle \mathbf{b}_{\omega,\rho}^{*}(\mathbf{X}_{-\omega}^{-1}), \mathbf{x}_0 \rangle \quad \text{almost surely,} \qquad \text{(B.3)}$$

Next, we apply Lemma B.1 for the function

$$f_i(\mathbf{x}_{-\infty}^{\infty}) = \log\langle \mathbf{h}^{(\omega,\rho)}(\mathbf{x}_{1-i}^{-1}), \mathbf{x}_0 \rangle = \log\langle \mathbf{b}^{(\omega,\rho)}(\mathbf{x}_{1-i}^{-1}, \mathbf{x}_{-\&}^{-1}), \mathbf{x}_0 \rangle$$

defined on $\mathbf{x}_{-\infty}^{\infty} = (\ldots, \mathbf{x}_{-1}, \mathbf{x}_0, \mathbf{x}_1)$. *Note that*

$$|f_i(\mathbf{X}_{-\infty}^{\infty})| = |\log\langle \mathbf{h}^{(\omega,\rho)}(\mathbf{X}_{1-i}^{-1}), \mathbf{x}_0 \rangle| \leq \sum_{j=1}^{d} |\log X_0^{(j)}|,$$

which has finite expectation, and

$$f_i(\mathbf{X}_{-\infty}^{\infty}) \to \langle b_{\omega,\rho}^{*}(X_{-\infty}^{-1}), X_0 \rangle \quad \text{almost surely as } i \to \infty$$

by Equation B.3. As $n \to \infty$, *Lemma B.1 yields*

$$
\begin{aligned}
W_n(\epsilon^{(\omega,\rho)}) \quad &= \tfrac{1}{n} \sum_{i=1}^{n} \log\langle \mathbf{h}^{(\omega,\rho)}(\mathbf{X}_1^{i-1}), \mathbf{X}_i \rangle \\
&= \tfrac{1}{n} \sum_{i=1}^{n} f_i(T^i X_{-\infty}^{\infty}) \\
&\to \mathbb{E}\{\log \mathbf{b}_{\omega,\rho}^{*}(X_{-\omega}^{-1}), X_0\} \\
&\overset{def}{=} \theta_{\omega,\rho} \quad \text{almost surely.}
\end{aligned}
$$

Therefore, by Equation B.1, we have

$$\liminf_{n\to\infty} W_n(\mathbf{B}) \geq \sup_{\omega,\rho} \theta_{\omega,\rho} \geq \sup_{\omega} \liminf_{\rho} \theta_{\omega,\rho} \quad \text{almost surely,}$$

and it suffices to show that the right-hand side is at least W^{*}.

(iii)   To this end, first, define, for Borel sets A, $B \subset \mathbb{R}_+^d$,

$$m_A(z) = \mathbb{P}\{\mathbf{X}_0 \in A | \mathbf{X}_{-\omega}^{-1} = z\}$$

and

$$\mu_\omega(B) = \mathbb{P}\{\mathbf{X}_{-\omega}^{-1} \in B\}.$$

Then, for any $\mathbf{s} \in$ support(μ_ω), and for all A,

$$\mathbb{P}_s^{*(\omega,\rho)}(A) = \mathbb{P}\{\mathbf{X}_0 \in A | \mathbb{E}\{(\mathbf{X}_{-\omega}^{-1} - \mathbf{s})^2\} \le 2\text{Var}(\mathbf{s})(1-\rho)\}$$

$$= \frac{\mathbb{P}\{\mathbf{X}_0 \in A, \mathbb{E}\{(\mathbf{X}_{-\omega}^{-1}-\mathbf{s})^2\} \le 2\text{Var}(\mathbf{s})(1-\rho)\}}{\mathbb{P}\{\mathbb{E}\{(\mathbf{X}_{-\omega}^{-1}-\mathbf{s})^2\} \le 2\text{Var}(\mathbf{s})(1-\rho)\}}$$

$$= \frac{1}{\mu_\omega(S_{s,2\text{Var}(\mathbf{s})(1-\rho)})} \int_{S_{s,2\text{Var}(\mathbf{s})(1-\rho)}} m_A(z)\mu_\omega(dz)$$

$$\to m_A(\mathbf{s}) = \mathbb{P}\{X_0 \in A | \mathbf{X}_{-\omega}^{-1} = \mathbf{s}\}$$

as $\rho \to 1$ and for μ_ω, almost all $\mathbf{s}$ by Lebesgue density theorem (see Lemma B.4), and therefore

$$\mathbb{P}_{\mathbf{X}_{-\omega}^{-1}}^{*(\omega,\rho)}(A) \to \mathbb{P}\{\mathbf{X}_0 \in A | \mathbf{X}_{-\omega}^{-1}\}$$

as $\rho \to 1$ for all A. Thus, using Lemma B.2 again, we have

$$\begin{aligned}
\liminf_\rho \theta_{\omega,\rho} &= \lim_\rho \theta_{\omega,\rho} \\
&= \lim_\rho \mathbb{E}\{\log \mathbf{b}_{\omega,\rho}^*(X_{-\omega}^{-1}), X_0\} \\
&= \mathbb{E}\{\log \langle \mathbf{b}_\omega^*(X_{-\omega}^{-1}), \mathbf{X}_0\rangle\} \\
&\quad \textit{(where } \mathbf{b}_\omega^*(\cdot) \textit{ is the log-optimum portfolio with respect} \\
&\quad \textit{to the conditional probability } \mathbb{P}\{\mathbf{X}_0 \in A | \mathbf{X}_{-\omega}^{-1}\}) \\
&= \mathbb{E}\{\max_{\mathbf{b}(\cdot)} \mathbb{E}\{\log \langle \mathbf{b}(\mathbf{X}_{-\omega}^{-1}), \mathbf{X}_0\rangle | \mathbf{X}_{-\omega}^{-1}\}\} \\
&= \mathbb{E}\{\mathbb{E}\{\log \langle \mathbf{b}_\omega^*(\mathbf{X}_{-\omega}^{-1}), \mathbf{X}_0\rangle | \mathbf{X}_{-\omega}^{-1}\}\} \\
&\stackrel{def}{=} \theta_\omega^*.
\end{aligned}$$

Next, to finish the proof, we appeal to the submartingale convergence theorem. First, note that the sequence

$$Y_\omega \stackrel{def}{=} \mathbb{E}\{\log \langle \mathbf{b}_\omega^*(\mathbf{X}_{-\omega}^{-1}), \mathbf{X}_0\rangle | \mathbf{X}_{-\omega}^{-1}\} = \max_{\mathbf{b}(\cdot)} \mathbb{E}\{\log \langle \mathbf{b}(\mathbf{X}_{-\omega}^{-1}), \mathbf{X}_0\rangle | \mathbf{X}_{-\omega}^{-1}\}$$

of random variables forms a submartingale, that is, $\mathbb{E}\{Y_{\omega+1} | Y_{-\omega}^{-1} \ge Y_\omega\}$. To see this, note that

$$\begin{aligned}
\mathbb{E}\{Y_{\omega+1} | \mathbf{X}_{-\omega}^{-1}\} &= \mathbb{E}\{\mathbb{E}\{\log \langle \mathbf{b}_{\omega+1}^*(\mathbf{X}_{-\omega-1}^{-1}), \mathbf{X}_0\rangle | \mathbf{X}_{-\omega-1}^{-1}\} | \mathbf{X}_{-\omega}^{-1}\} \\
&\ge \mathbb{E}\{\mathbb{E}\{\log \langle \mathbf{b}_\omega^*(\mathbf{X}_{-\omega}^{-1}), \mathbf{X}_0\rangle | \mathbf{X}_{-\omega-1}^{-1}\} | \mathbf{X}_{-\omega}^{-1}\} \\
&= \mathbb{E}\{\log \langle \mathbf{b}_\omega^*(\mathbf{X}_{-\omega}^{-1}), \mathbf{X}_0\rangle | \mathbf{X}_{-\omega-1}^{-1}\} \\
&= Y_\omega.
\end{aligned}$$

This sequence is bounded by

$$\max_{\mathbf{b}(\cdot)} \mathbb{E}\{\log\langle \mathbf{b}(\mathbf{X}_{-\infty}^{-1}), \mathbf{X}_0\rangle | \mathbf{X}_{-\infty}^{-1}\},$$

which has a finite expectation. The submartingale convergence theorem (see Stout 1974) implies that a submartingale is convergence almost surely, and $\sup_\omega \theta_\omega^*$ *is finite. In particular, by the submartingale property,* θ_ω^* *is a bounded increasing sequence, so that*

$$\sup_\omega \theta_\omega^* = \lim_{\omega\to\infty} \theta_\omega^*.$$

Applying Lemma B.3 with the σ-algebras

$$\sigma(\mathbf{X}_{-\omega}^{-1}) \nearrow \sigma(\mathbf{X}_{-\infty}^{-1})$$

yields

$$\sup_\omega \theta_\omega^* = \lim_{\omega\to\infty} \mathbb{E}\left\{\max_{\mathbf{b}(\cdot)} \mathbb{E}\{\log\langle \mathbf{b}(\mathbf{X}_{-\omega}^{-1}), \mathbf{X}_0\rangle | \mathbf{X}_{-\omega}^{-1}\}\right\}$$
$$= \mathbb{E}\left\{\max_{\mathbf{b}(\cdot)} \mathbb{E}\{\log\langle \mathbf{b}(\mathbf{X}_{-\infty}^{-1}), \mathbf{X}_0\rangle | \mathbf{X}_{-\infty}^{-1}\}\right\}$$
$$= W^*.$$

Then

$$\liminf_{n\to\infty} W_n(\mathbf{B}) \geq \sup_{\omega,\rho} \theta_{\omega,\rho} \geq \sup_\omega \liminf_\rho \theta_{\omega,\rho} = \sup_\omega \theta_\omega^* = W^* \quad almost\ surely,$$

and from the above three parts of proof, we can get that

$$\lim_{n\to\infty} \frac{1}{n}\log S_n(\mathbf{B}) = W^* \quad almost\ surely$$

and the proof of Theorem B.1 is finished.

B.2   Derivations of PAMR

B.2.1   Proof of Proposition 9.1

Proof   *First, if* $\ell_\epsilon^t = 0$, *then* $\mathbf{b}_t$ *satisfies the constraint and is clearly the optimal solution.*

To solve the problem in case of $\ell_\epsilon^t \neq 0$, *we define the Lagrangian for the optimization problem (9.2) as*

$$\mathcal{L}(\mathbf{b}, \tau, \lambda) = \frac{1}{2}\|\mathbf{b} - \mathbf{b}_t\|^2 + \tau(\mathbf{x}_t \cdot \mathbf{b} - \epsilon) + \lambda(\mathbf{b}\cdot\mathbf{1} - 1), \qquad (B.4)$$

where $\tau \geq 0$ *is a Lagrange multiplier related to the loss function,* λ *is a Lagrange multiplier associated with the simplex constraint, and* $\mathbf{1}$ *denotes a column vector of m 1s. Note that the nonnegativity of portfolio* $\mathbf{b}$ *is not considered, since introducing this term causes too much complexity, and alternatively we project the final portfolio into a simplex to enforce the constraint.*

Setting the partial derivatives of $\mathcal{L}$ with respect to $\mathbf{b}$ to zero gives

$$0 = \frac{\partial \mathcal{L}}{\partial \mathbf{b}} = (\mathbf{b} - \mathbf{b}_t) + \tau \mathbf{x}_t + \lambda \mathbf{1}.$$

Multiplying both sides by $\mathbf{1}^\top$, we can get $\lambda = -\tau \frac{\mathbf{x}_t \cdot \mathbf{1}}{m}$. Moreover, since $\bar{x}_t = \frac{\mathbf{x}_t \cdot \mathbf{1}}{m}$, where $\bar{x}_t$ is the mean of t-th price relatives, or the market return, we can rewrite λ as

$$\lambda = -\tau \bar{x}_t. \tag{B.5}$$

And the solution for $\mathcal{L}$ becomes

$$\mathbf{b} = \mathbf{b}_t - \tau(\mathbf{x}_t - \bar{x}_t \mathbf{1}). \tag{B.6}$$

Plugging Equation B.5 and Equation B.6 to Equation B.4, we get

$$\begin{aligned}
\mathcal{L}(\tau) &= \frac{1}{2}\tau^2 \|\mathbf{x}_t - \bar{x}_t \mathbf{1}\|^2 - \tau^2 \mathbf{x}_t \cdot (\mathbf{x}_t - \bar{x}_t \mathbf{1}) + \tau(\mathbf{b}_t \cdot \mathbf{x}_t - \epsilon) \\
&= -\frac{1}{2}\tau^2 \|\mathbf{x}_t - \bar{x}_t \mathbf{1}\|^2 + \tau(\mathbf{b}_t \cdot \mathbf{x}_t - \epsilon).
\end{aligned}$$

Note that in the above formula, we used the following formula:

$$\|\mathbf{x}_t - \bar{x}_t \mathbf{1}\|^2 = \mathbf{x}_t \cdot \mathbf{x}_t - 2\bar{x}_t(\mathbf{x}_t \cdot \mathbf{1}) + \bar{x}_t^2(\mathbf{1} \cdot \mathbf{1}) = \mathbf{x}_t \cdot \mathbf{x}_t - \bar{x}_t(\mathbf{x}_t \cdot \mathbf{1}) = \mathbf{x}_t \cdot (\mathbf{x}_t - \bar{x}_t \mathbf{1}).$$

Setting the derivative of $\mathcal{L}(\tau)$ with respect to τ to 0, we get

$$0 = \frac{\partial \mathcal{L}}{\partial \tau} = -\tau \|\mathbf{x}_t - \bar{x}_t \mathbf{1}\|^2 + \mathbf{b}_t \cdot \mathbf{x}_t - \epsilon.$$

Then τ can be set as

$$\tau = \frac{\mathbf{b}_t \cdot \mathbf{x}_t - \epsilon}{\|\mathbf{x}_t - \bar{x}_t \mathbf{1}\|^2}.$$

Since $\tau \geq 0$, we project τ to $[0, \infty)$; thus,

$$\tau = \max\left\{0, \frac{\mathbf{b}_t \cdot \mathbf{x}_t - \epsilon}{\|\mathbf{x}_t - \bar{x}_t \mathbf{1}\|^2}\right\} = \frac{\ell_\epsilon^t}{\|\mathbf{x}_t - \bar{x}_t \mathbf{1}\|^2}.$$

Note that in case of zero market volatility, that is, $\|\mathbf{x}_t - \bar{x}_t \mathbf{1}\|^2 = 0$, we just set $\tau = 0$. We can summarize the update scheme for the case of $\ell_\epsilon^t = 0$ and the case of $\ell_\epsilon^t > 0$ by setting τ. Thus, we simplify the notation following Equation 9.1 and show the unified update scheme.

B.2.2   Proof of Proposition 9.2

Proof   *We derive the solution of PAMR-1 following the same procedure as the derivation of PAMR. If the loss is nonzero, we get a Lagrangian*

$$\mathcal{L}(\mathbf{b}, \xi, \tau, \mu, \lambda) = \frac{1}{2}\|\mathbf{b} - \mathbf{b}_t\|^2 + \tau(\mathbf{x}_t \cdot \mathbf{b} - \epsilon) + \xi(C - \tau - \mu) + \lambda(\mathbf{1} \cdot \mathbf{b} - 1).$$

Setting the partial derivatives of $\mathcal{L}$ with respect to $\mathbf{b}$ to zero gives

$$0 = \frac{\partial \mathcal{L}}{\partial \mathbf{b}} = (\mathbf{b} - \mathbf{b}_t) + \tau \mathbf{x}_t + \lambda \mathbf{1},$$

Multiplying both sides by $\mathbf{1}^\top$, we can get $\lambda = -\tau \frac{\mathbf{x}_t \cdot \mathbf{1}}{m} = -\tau \bar{x}_t$. And the solution is

$$\mathbf{b} = \mathbf{b}_t - \tau(\mathbf{x}_t - \bar{x}_t \mathbf{1}).$$

Next, note that the minimum of the term $\xi(C - \tau - \mu)$ with respect to ξ is zero whenever $C - \tau - \mu = 0$. If $C - \tau - \mu \neq 0$, then the minimum can be made to approach $-\infty$. Since we need to maximize the dual, we can rule out the latter case and pose the following constraint on the dual variables, $C - \tau - \mu = 0$. The KKT conditions confine μ to be nonnegative, so we conclude that $\tau \leq C$. We can project τ to the interval $[0, C]$ and get

$$\tau = \max\left\{0, \min\left\{C, \frac{\mathbf{b}_t \cdot \mathbf{x}_t - \epsilon}{\|\mathbf{x}_t - \bar{x}_t \mathbf{1}\|^2}\right\}\right\} = \min\left\{C, \frac{\ell_\epsilon^t}{\|\mathbf{x}_t - \bar{x}_t \mathbf{1}\|^2}\right\}.$$

Again, we simplify the notation according to Equation 9.1 and show a unified update scheme.

B.2.3   Proof of Proposition 9.3

Proof   *We derive the solution similar to the derivations of PAMR and PAMR-1. In case that the loss is not 0, we can get the Lagrangian,*

$$\mathcal{L}(\mathbf{b}, \xi, \tau, \mu, \lambda) = \frac{1}{2}\|\mathbf{b} - \mathbf{b}_t\|^2 + \tau(\mathbf{b} \cdot \mathbf{x}_t - \epsilon) + C\xi^2 - \tau\xi + \lambda(\mathbf{1} \cdot \mathbf{b} - 1).$$

Setting the partial derivatives of $\mathcal{L}$ with respect to $\mathbf{b}$ to zero gives

$$0 = \frac{\partial \mathcal{L}}{\partial \mathbf{b}} = (\mathbf{b} - \mathbf{b}_t) + \tau \mathbf{x}_t + \lambda \mathbf{1},$$

Multiplying both sides by $\mathbf{1}^\top$, we can get $\lambda = -\tau \frac{\mathbf{x}_t \cdot \mathbf{1}}{m} = -\tau \bar{x}$. And the solution is

$$\mathbf{b} = \mathbf{b}_t - \tau(\mathbf{x}_t - \bar{x}\mathbf{1}).$$

Setting the partial derivatives of $\mathcal{L}$ with respect to ξ to zero gives

$$0 = \frac{\partial \mathcal{L}}{\partial \xi} = 2C\xi - \tau \implies \xi = \frac{\tau}{2C}.$$

Expressing ξ as above and replacing **b**, *we rewrite the Lagrangian as*

$$\tilde{\mathcal{L}}(\tau) = -\frac{\tau^2}{2}\left(\|\mathbf{x}_t - \bar{x}_t \mathbf{1}\|^2 + \frac{1}{2C}\right) + \tau(\mathbf{b}_t \cdot \mathbf{x}_t - \epsilon).$$

Taking the derivative with respect to τ and setting it to zero, we can get

$$0 = \frac{\partial \tilde{\mathcal{L}}}{\partial \tau} = -\tau\left(\|\mathbf{x}_t - \bar{x}_t \mathbf{1}\|^2 + \frac{1}{2C}\right) + (\mathbf{b}_t \cdot \mathbf{x}_t - \epsilon).$$

Then we get the update scheme of τ and project it to [0, ∞):

$$\tau = \max\left\{0, \frac{\mathbf{b}_t \cdot \mathbf{x}_t - \epsilon}{\|\mathbf{x}_t - \bar{x}_t \mathbf{1}\|^2 + \frac{1}{2C}}\right\} = \frac{\ell^t_\epsilon}{\|\mathbf{x}_t - \bar{x}_t \mathbf{1}\|^2 + \frac{1}{2C}}.$$

B.3 Derivations of CWMR

B.3.1 Proof of Proposition 10.1

Proof *Since considering the nonnegativity constraint introduces too much complexity, first we relax the optimization problem without it, and later we project the solution to the simplex domain to obtain the required portfolio.*

The Lagrangian for the optimization problem (10.3) *is*

$$\mathcal{L} = \frac{1}{2}\left(\log\left(\frac{det\Sigma_t}{det\Sigma}\right) + Tr(\Sigma_t^{-1}\Sigma) + (\mu_t - \mu)^\top \Sigma_t^{-1}(\mu_t - \mu)\right)$$
$$+ \lambda(\phi\mathbf{x}_t^\top \Sigma\mathbf{x}_t + \mu^\top\mathbf{x}_t - \epsilon) + \eta(\mu^\top\mathbf{1} - 1).$$

Taking the derivative of the Lagrangian with respect to μ and setting it to zero, we can get the update of μ

$$0 = \frac{\partial \mathcal{L}}{\partial \mu} = \Sigma_t^{-1}(\mu - \mu_t) + \lambda\mathbf{x}_t + \eta\mathbf{1} \implies \mu_{t+1} = \mu_t - \Sigma_t(\lambda\mathbf{x}_t + \eta\mathbf{1}), \quad \text{(B.7)}$$

where Σ_t *is assumed to be nonsingular. Multiplying both sides by* $\mathbf{1}^\top$, *we can get* η

$$1 = 1 - \mathbf{1}^\top\Sigma_t(\lambda\mathbf{x}_t + \eta\mathbf{1}) \implies \eta = -\lambda\bar{x}_t, \quad \text{(B.8)}$$

where $\bar{x}_t = \frac{\mathbf{1}^\top\Sigma_t\mathbf{x}_t}{\mathbf{1}^\top\Sigma_t\mathbf{1}}$ *denotes the confidence-weighted average of t-th price relatives.*

Plugging Equation B.8 to Equation B.7, we can get

$$\mu_{t+1} = \mu_t - \lambda\Sigma_t(\mathbf{x}_t - \bar{x}_t\mathbf{1}). \quad \text{(B.9)}$$

Moreover, taking the derivative of the Lagrangian with respect to Σ *and setting it to zero, we can have the update of* Σ:

$$0 = \frac{\partial\mathcal{L}}{\partial\Sigma} = -\frac{1}{2}\Sigma^{-1} + \frac{1}{2}\Sigma_t^{-1} + \lambda\phi\mathbf{x}_t\mathbf{x}_t^\top \implies \Sigma_{t+1}^{-1} = \Sigma_t^{-1} + 2\lambda\phi\mathbf{x}_t\mathbf{x}_t^\top.$$
$$\text{(B.10)}$$

Now let us solve the Lagrange multiplier λ_{t+1} using KKT conditions. First, following Dredze et al. (2008), we can compute the inverse using Woodbury identity (Golub and Van Loan 1996):

$$\Sigma_{t+1} = (\Sigma_t^{-1} + 2\lambda\phi\mathbf{x}_t\mathbf{x}_t^\top)^{-1} = \Sigma_t - \Sigma_t\mathbf{x}_t\frac{2\lambda\phi}{1 + 2\lambda\phi\mathbf{x}_t^\top\Sigma_t\mathbf{x}_t}\mathbf{x}_t^\top\Sigma_t. \quad \text{(B.11)}$$

The KKT conditions imply that either $\lambda = 0$, and no update is needed; or the constraint in the optimization problem (10.3) is an equality after the update. Taking Equation B.9 and Equation B.11 to the equality version of the first constraint, we can get

$$\epsilon - (\mu_t - \lambda\Sigma_t(\mathbf{x}_t - \bar{x}_t\mathbf{1})) \cdot \mathbf{x}_t = \phi\left(\mathbf{x}_t^\top\left(\Sigma_t - \Sigma_t\mathbf{x}_t\frac{2\lambda\phi}{1 + 2\lambda\phi\mathbf{x}_t^\top\Sigma_t\mathbf{x}_t}\mathbf{x}_t^\top\Sigma_t\right)\mathbf{x}_t\right).$$

Let $M_t = \mu_t^\top\mathbf{x}_t$ be the return mean, $V_t = \mathbf{x}_t^\top\Sigma_t\mathbf{x}_t$ be the return variance of the t-th trading period before updating, and $W_t = \mathbf{x}_t^\top\Sigma_t\mathbf{1}$ be the return variance of the t-th price relative with cash. We can simplify the preceding equation to

$$\lambda^2(2\phi V_t^2 - 2\phi\bar{x}_t V_t W_t) + \lambda(2\phi\epsilon V_t - 2\phi V_t M_t + V_t - \bar{x}_t W_t) + (\epsilon - M_t - \phi V_t) = 0. \quad \text{(B.12)}$$

Let us define $a = 2\phi V_t^2 - 2\phi\bar{x}_t V_t W_t$, $b = 2\phi\epsilon V_t - 2\phi V_t M_t + V_t - \bar{x}_t W_t$, and $c = \epsilon - M_t - \phi V_t$. Note that the above quadratic form equation may have two, one, or zero real roots. We can calculate its real roots (two real roots case: γ_{t1} and γ_{t2}; one real root case: γ_{t3}) as follows:

$$\gamma_{t1} = \frac{-b + \sqrt{b^2 - 4ac}}{2a}, \quad \gamma_{t2} = \frac{-b - \sqrt{b^2 - 4ac}}{2a}, \quad or \quad \gamma_{t3} = -\frac{c}{b}.$$

To ensure the nonnegativity of the Lagrangian multiplier, we can project its value to $[0, +\infty)$:

$$\lambda = \max\{\gamma_{t1}, \gamma_{t2}, 0\}, \quad or \quad \lambda = \max\{\gamma_{t3}, 0\}, \quad or \quad \lambda = 0.$$

Note that the above equations, respectively, correspond to three cases of real roots (two, one, or zero).

In practical computation, as we only adopt the diagonal elements of a covariance matrix, it is equivalent to compute λ from Equation B.12 but update the covariance matrix with the following rule instead of Equation B.10:

$$\Sigma_{t+1}^{-1} = \Sigma_t^{-1} + 2\lambda\phi diag^2(\mathbf{x}_t),$$

where $diag(\mathbf{x}_t)$ denotes a diagonal matrix with the elements of $\mathbf{x}_t$ on its main diagonal.

B.3.2   Proof of Proposition 10.2

Proof   *Similar to the proof of Proposition 10.1, we relax the optimization problem without the nonnegativity constraint and project the solution to the simplex domain to obtain the required portfolio.*

The Lagrangian for the optimization problem (10.4) is

$$\mathcal{L} = \frac{1}{2}\left(\log\left(\frac{det\Upsilon_t^2}{det\Upsilon^2}\right) + Tr(\Upsilon_t^{-2}\Upsilon^2) + (\mu_t - \mu)^\top\Upsilon_t^{-2}(\mu_t - \mu)\right)$$
$$+ \lambda(\phi\|\Upsilon\mathbf{x}_t\| + \mu^\top\mathbf{x}_t - \epsilon) + \eta(\mu^\top\mathbf{1} - 1).$$

Taking the derivative of the Lagrangian with respect to μ and setting it to zero, we can get the update of μ,

$$0 = \frac{\partial\mathcal{L}}{\partial\mu} = \Upsilon_t^{-2}(\mu - \mu_t) + \lambda\mathbf{x}_t + \eta\mathbf{1} \implies \mu_{t+1} = \mu_t - \Upsilon_t^2(\lambda\mathbf{x}_t + \eta\mathbf{1}),$$

where Υ_t is nonsingular. Multiplying both sides by $\mathbf{1}^\top$, we can get

$$1 = 1 - \mathbf{1}^\top\Upsilon_t^2(\lambda\mathbf{x}_t + \eta\mathbf{1}) \implies \eta = -\lambda\bar{x}_t,$$

where $\bar{x}_t = \frac{\mathbf{1}^\top\Upsilon_t^2\mathbf{x}_t}{\mathbf{1}^\top\Upsilon_t^2\mathbf{1}}$ is the confidence-weighted average of t-th price relatives. Plugging it into the update scheme of μ_{t+1}, we can get

$$\mu_{t+1} = \mu_t - \lambda\Upsilon_t^2(\mathbf{x}_t - \bar{x}_t\mathbf{1}).$$

Moreover, taking the derivative of the Lagrangian with respect to Υ and setting it to zero, we have

$$0 = \frac{\partial\mathcal{L}}{\partial\Upsilon} = -\Upsilon^{-1} + \frac{1}{2}\Upsilon_t^{-2}\Upsilon + \frac{1}{2}\Upsilon\Upsilon_t^{-2} + \lambda\phi\frac{\mathbf{x}_t\mathbf{x}_t^\top\Upsilon}{2\sqrt{\mathbf{x}_t^\top\Upsilon^2\mathbf{x}_t}} + \lambda\phi\frac{\Upsilon\mathbf{x}_t\mathbf{x}_t^\top}{2\sqrt{\mathbf{x}_t^\top\Upsilon^2\mathbf{x}_t}}.$$

We can solve the preceding equation to obtain Υ^{-2}:

$$\Upsilon_{t+1}^{-2} = \Upsilon_t^{-2} + \lambda\phi\frac{\mathbf{x}_t\mathbf{x}_t^\top}{\sqrt{\mathbf{x}_t^\top\Upsilon_{t+1}^2\mathbf{x}_t}}.$$

The preceding two updates can be expressed in terms of the covariance matrix,

$$\mu_{t+1} = \mu_t - \lambda\Sigma_t(\mathbf{x}_t - \bar{x}_t\mathbf{1}), \quad \Sigma_{t+1}^{-1} = \Sigma_t^{-1} + \lambda\phi\frac{\mathbf{x}_t\mathbf{x}_t^\top}{\sqrt{\mathbf{x}_t^\top\Sigma_{t+1}\mathbf{x}_t}}. \quad (B.13)$$

Here, Σ_{t+1} is positive semidefinite (PSD) and nonsingular.

Now, let us solve the Lagrangian multiplier using its KKT condition. Following Crammer et al. (2008), we compute the inverse using Woodbury identity (Golub and Van Loan 1996):

$$\Sigma_{t+1} = \Sigma_t - \Sigma_t\mathbf{x}_t\left(\frac{\lambda\phi}{\sqrt{\mathbf{x}_t^\top\Sigma_{t+1}\mathbf{x}_t} + \lambda\phi\mathbf{x}_t^\top\Sigma_t\mathbf{x}_t}\right)\mathbf{x}_t^\top\Sigma_t. \quad (B.14)$$

Similar to the proof of Proposition 10.1, we set $M_t = \boldsymbol{\mu}_t^\top \mathbf{x}_t$, $V_t = \mathbf{x}_t^\top \boldsymbol{\Sigma}_t \mathbf{x}_t$, $W_t = \mathbf{x}_t^\top \boldsymbol{\Sigma}_t \mathbf{1}$, and $U_t = \mathbf{x}_t^\top \boldsymbol{\Sigma}_{t+1} \mathbf{x}_t$. Multiplying the preceding equation by $\mathbf{x}_t^\top$ (left) and $\mathbf{x}_t$ (right), we get $U_t = V_t - V_t \left(\frac{\lambda\phi}{\sqrt{U_t} + \lambda\phi V_t} \right) V_t$, which can be solved for U_t:

$$\sqrt{U_t} = \frac{-\lambda\phi V_t + \sqrt{\lambda^2\phi^2 V_t^2 + 4V_t}}{2}. \tag{B.15}$$

The KKT condition implies that either $\lambda = 0$, and no update is needed; or the constraint in the optimization problem (10.4) is an equality after the update. Substituting Equations B.13 and B.15 into the equality version of the constraint, after rearranging in terms of λ, we get

$$\lambda^2 \left(\left(V_t - \bar{x}_t W_t + \frac{\phi^2 V_t}{2} \right)^2 - \frac{\phi^4 V_t^2}{4} \right) + 2\lambda(\epsilon - M_t) \left(V_t - \bar{x}_t W_t + \frac{\phi^2 V_t}{2} \right) \tag{B.16}$$

$$+ (\epsilon - M_t)^2 - \phi^2 V_t = 0.$$

Let $a = \left(V_t - \bar{x}_t W_t + \frac{\phi^2 V_t}{2} \right)^2 - \frac{\phi^4 V_t^2}{4}$, $b = 2(\epsilon - M_t)\left(V_t - \bar{x}_t W_t + \frac{\phi^2 V_t}{2} \right)$, and $c = (\epsilon - M_t)^2 - \phi^2 V_t$. Note that we only consider real roots of the quadratic form equation. Thus, we can obtain γ_t as its roots (two real roots case: γ_{t1} and γ_{t2}; one real root case: γ_{t3}):

$$\gamma_{t1} = \frac{-b + \sqrt{b^2 - 4ac}}{2a}, \quad \gamma_{t2} = \frac{-b - \sqrt{b^2 - 4ac}}{2a} \quad or \quad \gamma_{t3} = -\frac{c}{b}.$$

To ensure the nonnegativity of the Lagrangian multiplier, we project the roots to $[0, +\infty)$:

$$\lambda = \max\{\gamma_{t1}, \gamma_{t2}, 0\}, \quad or \quad \lambda = \max\{\gamma_{t3}, 0\}, \quad or \quad \lambda = 0,$$

which corresponds to three cases (two, one, or zero real roots), respectively.

Following the Proof of Proposition 10.1, we can update the diagonal covariance matrix as

$$\boldsymbol{\Sigma}_{t+1}^{-1} = \boldsymbol{\Sigma}_t^{-1} + \lambda \frac{\phi}{\sqrt{U_t}} diag^2(\mathbf{x}_t),$$

where $diag(\mathbf{x}_t)$ denotes the diagonal matrix with the elements of $\mathbf{x}_t$ on its main diagonal.

B.4   Derivation of OLMAR

B.4.1   Proof of Proposition 11.1

Proof *Since introducing a nonnegative constraint of the simplex constraint causes much difficulty (Helmbold et al. 1998), first we do not consider it and finally project on the simplex domain.*

The Lagrangian of the optimization problem OLMAR is

$$\mathcal{L}(\mathbf{b}, \lambda, \eta) = \frac{1}{2}\|\mathbf{b} - \mathbf{b}_t\|^2 + \lambda(\epsilon - \mathbf{b} \cdot \tilde{\mathbf{x}}_{t+1}) + \eta(\mathbf{b} \cdot \mathbf{1} - 1),$$

where $\lambda \geq 0$ and η are the Lagrangian multipliers. Taking the gradient with respect to $\mathbf{b}$ and setting it to zero, we get

$$0 = \frac{\partial \mathcal{L}}{\partial \mathbf{b}} = (\mathbf{b} - \mathbf{b}_t) - \lambda \tilde{\mathbf{x}}_{t+1} + \eta \mathbf{1} \implies \mathbf{b} = \mathbf{b}_t + \lambda \tilde{\mathbf{x}}_{t+1} - \eta \mathbf{1},$$

Multiplying both sides by $\mathbf{1}^\top$, we get

$$1 = 1 + \lambda \tilde{\mathbf{x}}_{t+1} \cdot \mathbf{1} - \eta m \implies \eta = \lambda \bar{x}_{t+1},$$

where $\bar{x}_{t+1}$ denotes the average predicted price relative (market). Plugging the above equation to the update of $\mathbf{b}$, we get the update of $\mathbf{b}$,

$$\mathbf{b} = \mathbf{b}_t + \lambda(\tilde{\mathbf{x}}_{t+1} - \bar{x}_{t+1}\mathbf{1}),$$

To solve the Lagrangian multiplier, let us plug the above equation to the Lagrangian,

$$\mathcal{L}(\lambda) = \lambda(\epsilon - \mathbf{b}_t \cdot \tilde{\mathbf{x}}_{t+1}) - \frac{1}{2}\lambda^2\|\tilde{\mathbf{x}}_{t+1} - \bar{x}_{t+1}\mathbf{1}\|^2$$

Taking derivative with respect to λ and setting it to zero, we get

$$0 = \frac{\partial \mathcal{L}}{\partial \lambda} = (\epsilon - \mathbf{b}_t \cdot \tilde{\mathbf{x}}_{t+1}) - \lambda\|\tilde{\mathbf{x}}_{t+1} - \bar{x}_{t+1}\mathbf{1}\|^2 \implies \lambda = \frac{\epsilon - \mathbf{b}_t \cdot \tilde{\mathbf{x}}_{t+1}}{\|\tilde{\mathbf{x}}_{t+1} - \bar{x}_{t+1}\mathbf{1}\|^2}.$$

Further projecting λ to $[0, +\infty)$, we get $\lambda = \max\left\{0, \frac{\epsilon - \mathbf{b}_t \cdot \tilde{\mathbf{x}}_{t+1}}{\|\tilde{\mathbf{x}}_{t+1} - \bar{x}_{t+1}\mathbf{1}\|^2}\right\}$.

Appendix C

Supplementary Data and Portfolio Statistics

This section provides some supplementary data and portfolio statistics, which mainly complement the observations in Section 13.6.

Similar to Table 13.5, Tables C.1, C.2, C.3, and C.4 show some descriptive statistics on the NYSE (N) dataset, SP500 dataset, MSCI dataset, and DJIA dataset, respectively.

Table C.5 illustrates the top five average allocation weights of the proposed strategies on the datasets except NYSE (O). Connecting these weights with the descriptive statistics, we can have similar observations as that of NYSE (O). That is, the proposed algorithms put more weights on the volatile assets so as to exploit the volatility of the assets, and most of the top average allocation weights of the mean reversion algorithms are assets with negative autocorrelations, except DJIA.

Table C.1 *Some descriptive statistics on the NYSE (N) dataset*

Stat.	1	2	3	4	5	6	7	8
Cum	15.0504	3.7815	9.9188	**32.7579**	13.0692	9.1587	10.2382	9.3461
Mean	1.0006	1.0005	1.0005	1.0007	1.0007	1.0006	1.0005	1.0007
Std	0.0184	0.0237	0.0174	0.0163	0.0243	0.0203	0.0180	**0.0251**
Ac	−0.0084	0.0321	0.0365	−0.0140	−0.0098	−0.0134	−0.0104	0.0060

Stat.	9	10	11	12	13	14	15	16
Cum	12.3507	0.0407	16.8484	6.8311	19.7812	**43.8699**	19.6711	2.0467
Mean	1.0006	1.0002	1.0007	1.0005	1.0007	1.0007	1.0006	1.0010
Std	0.0185	**0.0356**	**0.0246**	0.0183	0.0221	0.0153	0.0157	**0.0436**
Ac	−0.0193	0.0291	−0.0177	−0.0266	0.0145	0.0149	**−0.0507**	**−0.0963**

Stat.	17	18	19	20	21	22	23
Cum	0.3934	**28.7380**	17.2551	**83.5067**	17.4584	10.4056	**32.7829**
Mean	1.0002	1.0007	1.0006	1.0009	1.0008	1.0006	1.0007
Std	0.0245	0.0179	0.0154	0.0181	**0.0251**	0.0226	0.0191
Ac	0.0206	0.0139	**−0.0368**	−0.0192	0.0055	**−0.0339**	−0.0478

Note: The top five of each statistics are highlighted. Each column denotes the index of an asset.

Table C.2 *Some descriptive statistics on the SP500 dataset*

Stat.	1	2	3	4	5	6	7	8
Cum	0.9381	1.4478	**2.4279**	1.1038	1.2034	1.2841	**1.6480**	1.3822
Mean	1.0002	1.0007	1.0010	1.0002	1.0004	1.0006	1.0006	1.0005
Std	0.0229	0.0280	0.0245	0.0180	0.0233	0.0273	0.0181	0.0223
Ac	−0.0090	−0.0103	0.0032	−0.0363	0.0256	0.0033	0.0783	0.0510
	9	10	11	12	13	14	15	16
Cum	1.4807	1.0293	1.0572	0.8625	1.1531	0.6044	1.3816	0.8419
Mean	1.0006	1.0002	1.0003	1.0005	1.0004	0.9998	1.0010	1.0001
Std	0.0255	0.0206	0.0215	**0.0363**	0.0247	0.0202	**0.0392**	0.0234
Ac	−0.0389	0.0590	−0.0182	−0.0386	0.0063	0.0508	−0.0618	**−0.0962**
	17	18	19	20	21	22	23	24
Cum	1.2405	**3.7792**	**2.2260**	1.1244	0.6523	1.1119	0.8263	**1.8606**
Mean	1.0004	1.0016	1.0013	1.0003	1.0000	1.0003	1.0000	1.0009
Std	0.0210	**0.0323**	**0.0369**	0.0213	0.0244	0.0212	0.0176	**0.0296**
Ac	−0.0206	**−0.0431**	0.0094	**−0.0755**	−0.0150	0.0017	**−0.0401**	0.0229
	25							
Cum	0.8738							
Mean	1.0002							
Std	0.0242							
Ac	0.0299							

Note: The top five of each statistics are highlighted. Each column denotes the index of an asset.

Table C.3 Some descriptive statistics on the MSCI dataset

Stat.	1	2	3	4	5	6	7	8
Cum	0.9095	**1.2541**	0.8248	1.0638	1.0133	1.0242	**1.3189**	0.8063
Mean	1.0000	1.0005	1.0000	1.0002	1.0003	1.0002	1.0004	0.9999
Std	0.0164	0.0219	0.0190	0.0152	**0.0242**	0.0198	0.0180	0.0162
Ac	**−0.1305**	−0.0099	**−0.0426**	−0.0240	−0.0105	−0.0072	0.0082	0.0323

	9	10	11	12	13	14	15	16
Cum	0.8742	0.7265	0.9160	0.6586	**1.5040**	0.2894	**1.2153**	0.5191
Mean	1.0003	0.9999	1.0001	0.9998	1.0005	0.9991	1.0003	0.9996
Std	**0.0281**	0.0178	0.0187	0.0196	0.0125	**0.0256**	0.0177	**0.0232**
Ac	−0.0322	0.0415	−0.0225	−0.0103	0.0237	0.0188	−0.0257	0.0624

	17	18	19	20	21	22	23	24
Cum	0.9653	0.8714	0.8123	**1.2018**	1.1802	0.5670	0.5080	0.7285
Mean	1.0001	1.0001	1.0000	1.0004	1.0004	0.9996	0.9997	0.9998
Std	0.0192	0.0195	0.0220	0.0191	0.0202	0.0196	**0.0250**	0.0170
Ac	−0.0228	**−0.0549**	**−0.0324**	0.0302	0.0275	0.0600	0.0410	**−0.1259**

Note: The top five of each statistics are highlighted. Each column denotes the index of an asset.

Table C.4 *Some descriptive statistics on the DJIA dataset*

Stat.	1	2	3	4	5	6	7	8
Cum	0.7085	0.5378	**1.1414**	**1.1884**	0.6870	0.7389	0.5449	**1.1572**
Mean	0.9997	0.9991	1.0004	1.0007	0.9996	0.9997	0.9993	1.0004
Std	0.0263	0.0260	0.0174	0.0269	0.0278	0.0256	**0.0314**	0.0161
Ac	−0.0315	0.0094	0.0230	**−0.0691**	0.0270	**−0.0961**	−0.0479	**−0.0895**

Stat.	9	10	11	12	13	14	15	16
Cum	0.5459	0.4287	0.7718	0.5806	0.6904	0.5323	0.5034	0.3608
Mean	0.9991	0.9988	0.9996	0.9992	0.9996	0.9992	0.9988	0.9990
Std	0.0256	0.0291	0.0166	0.0248	0.0268	**0.0309**	0.0198	**0.0398**
Ac	0.0597	0.0011	0.0280	−0.0360	−0.0066	−0.0033	0.0240	0.0041

Stat.	17	18	19	20	21	22	23	24
Cum	**1.0191**	0.6048	**1.0863**	0.8560	0.9178	0.9361	0.9787	0.8797
Mean	1.0003	0.9996	1.0003	1.0001	1.0001	1.0002	1.0002	1.0000
Std	0.0223	**0.0357**	0.0180	0.0264	0.0212	0.0249	0.0208	0.0202
Ac	−0.0506	0.0208	−0.0622	**−0.0963**	**−0.0658**	0.0257	−0.0330	−0.0578

Stat.	25	26	27	28	29	30
Cum	0.5947	0.5197	0.6710	0.8309	0.9872	0.9311
Mean	0.9994	0.9995	0.9994	0.9998	1.0003	1.0001
Std	0.0281	**0.0398**	0.0197	0.0179	0.0238	0.0209
Ac	0.0426	−0.0154	0.0873	−0.0434	0.0335	0.0288

Note: The top five of each statistics are highlighted. Each column denotes the index of an asset.

Table C.5 The top five (average) allocation weights of the proposed strategies on five datasets

NYSE (N)

Asset #	16	10	5	17	11	Asset #	13	4	16	23	20
Anticor	0.15	0.08	0.06	0.06	0.06	CORN	0.15	0.15	0.15	0.07	0.05
Asset #	16	11	5	10	22	Asset #	16	11	10	22	5
PAMR	0.18	0.07	0.06	0.06	0.06	OLMAR	0.18	0.07	0.07	0.06	0.06

TSE

Asset #	18	24	71	79	74	Asset #	51	87	25	71	18
Anticor	0.09	0.08	0.07	0.06	0.06	CORN	0.14	0.13	0.12	0.10	0.05
Asset #	24	18	71	79	74	Asset #	24	18	71	79	32
PAMR	0.12	0.08	0.07	0.04	0.04	OLMAR	0.10	0.09	0.08	0.05	0.04

SP500

Asset #	15	12	19	25	21	Asset #	19	24	15	18	8
Anticor	0.09	0.08	0.07	0.06	0.05	CORN	0.20	0.14	0.14	0.13	0.08
Asset #	15	19	12	18	24	Asset #	15	12	19	24	18
PAMR	0.10	0.08	0.07	0.07	0.06	OLMAR	0.09	0.08	0.07	0.06	0.06

MSCI

Asset #	9	2	7	1	20	Asset #	2	24	1	7	6
Anticor	0.17	0.16	0.14	0.07	0.06	CORN	0.15	0.14	0.11	0.10	0.08
Asset #	24	14	9	16	10	Asset #	14	16	9	24	10
PAMR	0.11	0.10	0.09	0.09	0.07	OLMAR	0.12	0.10	0.09	0.09	0.08

DJIA

Asset #	16	18	26	14	7	Asset #	19	24	15	7	8
Anticor	0.08	0.07	0.07	0.06	0.05	CORN	0.20	0.14	0.14	0.10	0.08
Asset #	18	26	16	14	7	Asset #	26	10	16	18	7
PAMR	0.10	0.07	0.07	0.06	0.06	OLMAR	0.11	0.07	0.06	0.06	0.05

Note: "Asset #" denotes the indices of the allocated assets.

Bibliography

J. Abernethy, A. Agarwal, P. L. Barlett, and A. Rakhlin. A stochastic view of optimal regret through minimax duality. In *Proceedings of Annual Conference on Learning Theory*, Montreal, Quebec, 2009.

A. Agarwal, E. Hazan, S. Kale, and R. E. Schapire. Algorithms for portfolio management based on the newton method. In *Proceedings of International Conference on Machine Learning*, Pittsburgh, PA, 9–16, 2006.

A. Agarwal, P. Bartlett, and M. Dama. Optimal allocation strategies for the dark pool problem. In *Proceedings of International Conference on Artificial Intelligence and Statistics*, Chia Laguna Resort, Sardinia, 9–16, 2010.

R. Agrawal and R. Srikant. Mining sequential patterns. In *Proceedings of the Eleventh International Conference on Data Engineering*, Taipei, Taiwan, 3–14, 1995.

D. W. Aha. Case-based learning algorithms. In *Proceedings of the DARPA Case-Based Reasoning Workshop*, 147–158, 1991.

D. W. Aha, D. Kibler, and M. K. Albert. Instance-based learning algorithms. *Machine Learning*, 6(1):37–66, 1991.

K. Akcoglu, P. Drineas, and M.-Y. Kao. Fast universalization of investment strategies. *SIAM Journal on Computing*, 34(1):1–22, 2005.

I. Aldridge. *High-Frequency Trading: A Practical Guide to Algorithmic Strategies and Trading Systems*. Hoboken, NJ: Wiley, 2010.

P. Algoet. Universal schemes for prediction, gambling and portfolio selection. *The Annals of Probability*, 20(2):901–941, 1992.

P. Algoet and T. Cover. Asymptotic optimality asymptotic equipartition properties of log-optimum investments. *Annals of Probability*, 16:876–898, 1988.

P. H. Algoet. The strong law of large numbers for sequential decisions under uncertainty. *IEEE Transactions on Information Theory*, 40:609–633, 1994.

R. F. Almgren and N. Chriss. Optimal execution of portfolio transactions. *Journal of Risk*, 12:61–63, 2000.

F. R. Bach. Consistency of the group lasso and multiple kernel learning. *Journal of Machine Learning Research*, 9:1179–1225, 2008.

L. Bachelier. Théorie de la spéculation. *Annales Scientifiques de l'École Normale Supérieure*, 3(17):21–86, 1900.

P. Baldi and P. Brunak. *Bioinformatics: The Machine Learning Approach*, 2nd Edition. Cambridge, MA: MIT Press, 2001.

N. Barberis and R. Thaler. A survey of behavioural finance. In *Handbook of the Economics of Finance*, G. M. Constantinides, M. Harris, and R. Stulz (eds.), Elsevier, North Holland, Amsterdam, 1053–1128, 2003.

E. Bayraktar. Optimal trade execution in illiquid markets. *Mathematical Finance*, 21(4):681–701, 2011.

J. E. Beasley, N. Meade, and T. J. Chang. An evolutionary heuristic for the index tracking problem. *European Journal of Operational Research*, 148(3):621–643, 2003.

C. Y. Belentepe. *A Statistical View of Universal Portfolios*. PhD thesis, University of Pennsylvania, 2005.

D. J. Berndt and J. Clifford. Using dynamic time warping to find patterns in time series. In *KDD Workshop*, 359–370, 1994.

D. Bernoulli. Exposition of a new theory on the measurement of risk. *Econometrica*, 23:23–36, 1954.

D. Bertsimas and A. W. Lo. Optimal control of execution costs. *Journal of Financial Markets*, 1(1):1–50, 1998.

J. R. Birge and F. Louveaux. *Introduction to Stochastic Programming*. New York: Springer, 1997.

A. Blum and A. Kalai. Universal portfolios with and without transaction costs. *Machine Learning*, 35(3):193–205, 1999.

A. Blum and Y. Mansour. From external to internal regret. *Journal of Machine Learning Research*, 8:1307–1324, 2007.

W. F. M. D. Bondt and R. Thaler. Does the stock market overreact? *The Journal of Finance*, 40(3):793–805, 1985.

A. Borodin, R. El-Yaniv, and V. Gogan. Can we learn to beat the best stock. *Journal of Artificial Intelligence Research*, 21:579–594, 2004.

S. Boyd and L. Vandenberghe. *Convex Optimization*. New York: Cambridge University Press, 2004.

L. Breiman. The individual ergodic theorem of information theory. *The Annals of Mathematical Statistics*, 31:809–811, 1957 (Correction version 1960).

L. Breiman. Investment policies for expanding businesses optimal in a long-run sense. *Naval Research Logistics Quarterly*, 7(4):647–651, 1960.

L. Breiman. Optimal gambling systems for favorable games. *Proceedings of the Berkeley Symposium on Mathematical Statistics and Probability*, 1:65–78, 1961.

J. Brodie, I. Daubechies, C. De Mol, D. Giannone, and I. Loris. Sparse and stable Markowitz portfolios. *Proceedings of the National Academy of Sciences*, 106 (30):12267–12272, 2009.

N. A. Canakgoz and J. E. Beasley. Mixed-integer programming approaches for index tracking and enhanced indexation. *European Journal of Operational Research*, 196(1):384–399, 2009.

A. Cañete, J. Constanzo, and L. Salinas. Kernel price pattern trading. *Applied Intelligence*, 29(2):152–156, 2008.

L. J. Cao and F. E. H. Tay. Support vector machine with adaptive parameters in financial time series forecasting. *IEEE Transactions on Neural Networks*, 14(6): 1506–1518, 2003.

N. Cesa-Bianchi and G. Lugosi. *Prediction, Learning, and Games*. New York: Cambridge University Press, 2006.

E. Chan. *Quantitative Trading: How to Build Your Own Algorithmic Trading Business*. Hoboken, NJ: Wiley, 2008.

K. C. Chan. On the contrarian investment strategy. *The Journal of Business*, 61(2): 147–163, 1988.

L. K. C. Chan, N. Jegadeesh, and J. Lakonishok. Momentum strategies. *The Journal of Finance*, 51(5):1681–1713, 1996.

K. Chaudhuri and Y. Wu. Mean reversion in stock prices: Evidence from emerging markets. *Managerial Finance*, 29:22–37, 2003.

V. S. Cherkassky and F. Mulier. *Learning from Data: Concepts, Theory, and Methods*. New York: Wiley, 1998.

V. K. Chopra and W. T. Ziemba. The effect of errors in means, variances, and covariances on optimal portfolio choice. *The Journal of Portfolio Management*, 19: 6–11, 1993.

T. F. Coleman, Y. Li, and J. Henniger. Minimizing tracking error while restricting the number of assets. *Journal of Risk*, 8:33–56, 2006.

J. Conrad and G. Kaul. An anatomy of trading strategies. *Review of Financial Studies*, 11(3):489–519, 1998.

R. Cont. Empirical properties of asset returns: stylized facts and statistical issues. *Quantitative Finance*, 1(2):223–236, 2001.

P. Cootner. *The Random Character of Stock Market Prices*. Cambridge, MA: MIT Press, 1964.

T. Cover and E. Ordentlich. Universal portfolios with short sales and margin. In *Proceedings of Annual IEEE International Symposium on Information Theory*, Cambridge, MA, 174, 1998.

T. M. Cover. Universal portfolios. *Mathematical Finance*, 1(1):1–29, 1991.

T. M. Cover. Universal data compression and portfolio selection. In *Proceedings of Annual IEEE Symposium on Foundations of Computer Science*, Burlington, VT, 534–538, 1996.

T. M. Cover and D. H. Gluss. Empirical Bayes stock market portfolios. *Advances in Applied Mathematics*, 7(2):170–181, 1986.

T. M. Cover and E. Ordentlich. Universal portfolios with side information. *IEEE Transactions on Information Theory*, 42(2):348–363, 1996.

T. M. Cover and J. A. Thomas. *Elements of Information Theory*. New York: Wiley, 1991.

K. Crammer, O. Dekel, J. Keshet, S. Shalev-Shwartz, and Y. Singer. Online passive–aggressive algorithms. *Journal of Machine Learning Research*, 7:551–585, 2006.

K. Crammer, M. Dredze, and A. Kulesza. Multi-class confidence weighted algorithms. In *Proceedings of the Conference on Empirical Methods in Natural Language Processing*, Singapore, 496–504, 2009.

K. Crammer, M. Dredze, and F. Pereira. Exact convex confidence-weighted learning. In *Proceedings of Annual Conference on Neural Information Processing Systems*, Vancouver, 345–352, 2008.

G. Creamer. *Using Boosting for Automated Planning and Trading Systems*. PhD thesis, Columbia University, 2007.

G. Creamer. Model calibration and automated trading agent for euro futures. *Quantitative Finance*, 12(4):531–545, 2012.

G. Creamer and S. Stolfo. A link mining algorithm for earnings forecast and trading. *Data Mining and Knowledge Discovery*, 18(3):419–445, 2009.

G. G. Creamer and Y. Freund. A boosting approach for automated trading. *Journal of Trading*, 2(3):84–96, 2007.

G. G. Creamer and Y. Freund. Automated trading with boosting and expert weighting. *Quantitative Finance*, 10(4):401–420, 2010.

J. E. Cross and A. R. Barron. Efficient universal portfolios for past-dependent target classes. *Mathematical Finance*, 13(2):245–276, 2003.

P. Das and A. Banerjee. Meta optimization and its application to portfolio selection. In *Proceedings of International Conference on Knowledge Discovery and Data Mining*, San Diego, 1163–1171, 2011.

V. DeMiguel, L. Garlappi, and R. Uppal. Optimal versus naive diversification: How inefficient is the $1 - n$ portfolio strategy? *Review of Financial Studies*, 22(5): 1915–1953, 2009.

M. A. H. Dempster, T. W. Payne, Y. Romahi, and G. W. P. Thompson. Computational learning techniques for intraday FX trading using popular technical indicators. *IEEE Transactions on Neural Networks*, 12(4):744–754, 2001.

E. Dimson. *Stock Market Anomalies*. Cambridge, MA: Cambridge University Press, 1988.

M. Dredze, K. Crammer, and F. Pereira. Confidence-weighted linear classification. In *Proceedings of International Conference on Machine Learning*, Helsinki, Finland, 246–271, 2008.

X. Du, R. Jin, L. Ding, V. E. Lee, and J. H. Thornton Jr. Migration motif: A spatial-temporal pattern mining approach for financial markets. In *Proceedings of the ACM SIGKDD International Conference on Knowledge Discovery and Data Mining*, Paris, France, 1135–1144, 2009.

J. Duchi, S. Shalev-Shwartz, Y. Singer, and T. Chandra. Efficient projections onto the l_1-ball for learning in high dimensions. In *Proceedings of International Conference on Machine Learning*, Helsinki, Finland, 272–279, 2008.

M. Durbin. *All About High-Frequency Trading*. New York: McGraw-Hill, 2010.

R. El-Yaniv. Competitive solutions for online financial problems. *ACM Computing Surveys*, 30:28–69, 1998.

J. Exley, S. Mehta, and A. Smith. Mean Reversion. Technical report, Faculty & Institute of Actuaries, Finance and Investment Conference, Brussels, 2004.

E. Fagiuoli, F. Stella, and A. Ventura. Constant rebalanced portfolios and side-information. *Quantitative Finance*, 7(2):161–173, 2007.

E. F. Fama and K. R. French. The cross-section of expected stock returns. *The Journal of Finance*, 47(2):427–465, 1992.

M. Feder, N. Merhav, and M. Gutman. Universal prediction of individual sequences. *IEEE Transactions on Information Theory*, 38(4):1258–1270, 1992.

M. Finkelstein and R. Whitley. Optimal strategies for repeated games. *Advances in Applied Probability*, 13(2):415–428, 1981.

T. Foucault, O. Kadan, and E. Kandel. Limit order book as a market for liquidity. *Review of Financial Studies*, 18(4):1171–1217, 2005.

W. J. Fu. Penalized regressions: The bridge versus the lasso. *Journal of Computational and Graphical Statistics*, 7(3):397–416, 1998.

A. A. Gaivoronski and F. Stella. Stochastic nonstationary optimization for finding universal portfolios. *Annals of Operations Research*, 100:165–188, 2000.

A. A. Gaivoronski and F. Stella. On-line portfolio selection using stochastic programming. *Journal of Economic Dynamics and Control*, 27(6):1013–1043, 2003.

K. Ganchev, Y. Nevmyvaka, M. Kearns, and J. W. Vaughan. Censored exploration and the dark pool problem. *Communications of the ACM*, 53(5):99–107, 2010.

M. Gilli and E. Këllezi. The threshold accepting heuristic for index tracking. In *Financial Engineering, E-Commerce, and Supply Chain*, P. M. Pardalos and V. Tsitsiringos (eds.), Boston: Kluwer Academic, 1–18, 2002.

G. H. Golub and C. F. Van Loan. *Matrix Computations*. Baltimore, MD: Johns Hopkins University Press, 1996.

T. F. Gosnell, A. J. Keown, and J. M. Pinkerton. The intraday speed of stock price adjustment to major dividend changes: Bid-ask bounce and order flow imbalances. *Journal of Banking & Finance*, 20(2):247–266, 1996.

R. Grinold and R. Kahn. *Active Portfolio Management: A Quantitative Approach for Producing Superior Returns and Controlling Risk*. New York: McGraw-Hill, 1999.

L. Györfi, G. Lugosi, and F. Udina. Nonparametric kernel-based sequential investment strategies. *Mathematical Finance*, 16(2):337–357, 2006.

L. Györfi, G. Ottucsák, and H. Walk. *Machine Learning for Financial Engineering*. Singapore: World Scientific, 2012.

L. Györfi, A. Urbán, and I. Vajda. Kernel-based semi-log-optimal empirical portfolio selection strategies. *International Journal of Theoretical and Applied Finance*, 10(3):505–516, 2007.

L. Györfi, F. Udina, and H. Walk. Nonparametric nearest neighbor based empirical portfolio selection strategies. *Statistics and Decisions*, 26(2):145–157, 2008.

L. Györfi and D. Schäfer. Nonparametric prediction. In *Advances in Learning Theory: Methods, Models and Applications*, J. Suykens, G. Horvath, and S. Basu (eds.), The Netherlands: IOS Press, Amsterdam, 339–354, 2003.

L. Györfi and I. Vajda. Growth optimal investment with transaction costs. In *Proceedings of the International Conference on Algorithmic Learning Theory*, Budapest, Hungary, 108–122, 2008.

L. Gyorfi and H. Walk. Empirical portfolio selection strategies with proportional transaction costs. *IEEE Transactions on Information Theory*, 58(10):6320–6331,
2012.

N. H. Hakansson. Optimal investment and consumption strategies under risk for
a class of utility functions. *Econometrica*, 38(5):587–607, 1970.

N. H. Hakansson. Capital growth and the mean-variance approach to portfolio
selection. *The Journal of Financial and Quantitative Analysis*, 6(1):517–557,
1971.

J. D. Hamilton. *Time Series Analysis*. Princeton, NJ: Princeton University Press,
1994.

J. D. Hamilton. Regime-switching models. In *New Palgrave Dictionary of Economics*,
S. N. Durlauf and L. E. Blume (eds.), New York: Palgrave McMillan, 53–57,
2008.

M. R. Hardy. A regime-switching model of long-term stock returns. *North American
Actuarial Journal Society of Acutaries*, 5(2):41–53, 2001.

L. Harris. *Trading and Exchanges: Market Microstructure for Practitioners*. New
York: Oxford University Press, 2003.

C. R. Harvey, J. C. Liechty, M. W. Liechty, and P. Müller. Portfolio selection with
higher moments. *Quantitative Finance*, 10(5):469–485, 2010.

R. A. Haugen and J. Lakonishok. *The Incredible January Effect: The Stock Market's
Unsolved Mystery*. Homewood, IL: Dow Jones-Irwin, 1987.

E. Hazan. *Efficient Algorithms for Online Convex Optimization and Their Applications*. PhD thesis, Princeton University, 2006.

E. Hazan, A. Agarwal, and S. Kale. Logarithmic regret algorithms for online convex
optimization. *Machine Learning*, 69(2–3):169–192, 2007.

E. Hazan, A. Kalai, S. Kale, and A. Agarwal. Logarithmic regret algorithms for
online convex optimization. In *Proceedings of the Annual Conference on
Learning Theory*, 2006.

E. Hazan and S. Kale. On stochastic and worst-case models for investing. In *Proceedings of Annual Conference on Neural Information Processing Systems*,
Vancouver, 709–717, 2009.

E. Hazan and S. Kale. An online portfolio selection algorithm with regret logarithmic in price variation. *Mathematical Finance*, 25(2):288–310, 2015.

E. Hazan and C. Seshadhri. Efficient learning algorithms for changing environments.
In *Proceedings of the International Conference on Machine Learning*, Montreal,
393–400, 2009.

D. P. Helmbold, R. E. Schapire, Y. Singer, and M. K. Warmuth. On-line portfolio selection using multiplicative updates. In *Proceedings of the International
Conference on Machine Learning*, Bari, Italy, 243–251, 1996.

D. P. Helmbold, R. E. Schapire, Y. Singer, and M. K. Warmuth. A comparison of new
and old algorithms for a mixture estimation problem. *Machine Learning*, 27(1):
97–119, 1997.

D. P. Helmbold, R. E. Schapire, Y. Singer, and M. K. Warmuth. On-line portfolio
selection using multiplicative updates. *Mathematical Finance*, 8(4):325–347,
1998.

M. Herbster and M. K. Warmuth. Tracking the best expert. *Machine Learning*, 32(2): 151–178, 1998.

E. Hillebrand. *Mean Reversion Models of Financial Markets*. PhD thesis, University of Bremen, 2003.

D. Huang, J. Zhou, B. Li, S. C. Hoi, and S. Zhou. Robust median reversion strategy for on-line portfolio selection. In *Proceedings of the Twenty-Third International Joint Conference on Artificial Intelligence*, 2006–2012, AAAI Press, Beijing, China, 2013.

S.-H. Huang, S.-H. Lai, and S.-H. Tai. A learning-based contrarian trading strategy via a dual-classifier model. *ACM Transactions on Interactive Intelligent Systems*, 2:20:1–20:20, 2011.

J. C. Hull. *Options, Futures, and Other Derivatives*. Upper Saddle River, NJ: Prentice Hall, 1997.

G. Iyengar. Universal investment in markets with transaction costs. *Mathematical Finance*, 15(2):359–371, 2005.

F. Jamshidian. Asymptotically optimal portfolios. *Mathematical Finance*, 2(2):131–150, 1992.

N. Jegadeesh. Evidence of predictable behavior of security returns. *Journal of Finance*, 45(3):881–898, 1990.

A. Kalai and S. Vempala. Efficient algorithms for universal portfolios. *Journal of Machine Learning Research*, 3:423–440, 2002.

J. O. Katz and D. L. McCormick. *The Encyclopedia of Trading Strategies*. New York: McGraw-Hill, 2000.

M. Kearns, A. Kulesza, and Y. Nevmyvaka. Empirical limitations on high frequency trading profitability. *Journal of Trading*, 5(4):50–62, 2010.

D. B. Keim and A. Madhavan. Anatomy of the trading process empirical evidence on the behavior of institutional traders. *Journal of Financial Economics*, 37(3): 371–398, 1995.

J. Kelly. A new interpretation of information rate. *Bell Systems Technical Journal*, 35:917–926, 1956.

E. Keogh. Exact indexing of dynamic time warping. In *Proceedings of the 28th International Conference on Very Large Data Bases*, 406–417, 2002.

E. J. Keogh and M. J. Pazzani. Scaling up dynamic time warping for datamining applications. In *Proceedings of the Sixth ACM SIGKDD International Conference on Knowledge Discovery and Data Mining*, Boston, MA, 285–289, 2000.

T. Kimoto, K. Asakawa, M. Yoda, and M. Takeoka. Stock market prediction system with modular neural networks. *Neural Networks in Finance and Investing*, 343–357, 1993.

R. Kissell, M. Glantz, and R. Malamut. *Optimal Trading Strategies: Quantitative Approaches for Managing Market Impact and Trading Risk*. New York: AMACOM, 2003.

W. M. Koolen and V. Vovk. Buy low, sell high. In *Proceedings of International Conference on Algorithmic Learning Theory*, Lyon, France, 335–349, 2012.

S. S. Kozat and A. C. Singer. Universal constant rebalanced portfolios with switching. In *Proceedings of the International Conference on Acoustics, Speech, and Signal Processing*, Honolulu, 1129–1132, 2007.

S. S. Kozat and A. C. Singer. Universal switching portfolios under transaction costs. In *Proceedings of the International Conference on Acoustics, Speech, and Signal Processing*, Las Vegas, NV, 5404–5407, 2008.

S. S. Kozat and A. C. Singer. Switching strategies for sequential decision problems with multiplicative loss with application to portfolios. *IEEE Transactions on Signal Processing*, 57(6):2192–2208, 2009.

S. S. Kozat and A. C. Singer. Universal randomized switching. *IEEE Transactions on Signal Processing*, 58:3, 2010.

S. S. Kozat and A. C. Singer. Universal semiconstant rebalanced portfolios. *Mathematical Finance*, 21(2):293–311, 2011.

S. S. Kozat, A. C. Singer, and A. J. Bean. Universal portfolios via context trees. In *Proceedings of the International Conference on Acoustics, Speech, and Signal Processing*, Las Vegas, NV, 2093–2096, 2008.

S. S. Kozat, A. C. Singer, and A. J. Bean. A tree-weighting approach to sequential decision problems with multiplicative loss. *Signal Processing*, 91(4):890–905, 2011.

S. Kullback and R. Leibler. On information and sufficiency. *Annals of Mathematical Statistics*, 22:79–86, 1951.

H. A. Latané. Criteria for choice among risky ventures. *The Journal of Political Economy*, 67(2):144–155, 1959.

T. Levina and G. Shafer. Portfolio selection and online learning. *International Journal of Uncertainty, Fuzziness and Knowledge-Based Systems*, 16(4):437–473, 2008.

B. Li and S. C. Hoi. Online portfolio selection: A survey. *ACM Computing Surveys*, 36:35:1–35:36, 2014.

B. Li, S. C. Hoi, and V. Gopalkrishnan. CORN: Correlation-driven nonparametric learning approach for portfolio selection. *ACM Transactions on Intelligent Systems and Technology*, 2(3):21:1–21:29, 2011a.

B. Li, S. C. Hoi, D. Sahoo, and Z. Liu. Moving average reversion strategy for on-line portfolio selection. *Artificial Intelligence*, 222:104–123, 2015.

B. Li, S. C. Hoi, P. Zhao, and V. Gopalkrishnan. Confidence weighted mean reversion strategy for on-line portfolio selection. In *Proceedings of the International Conference on Artificial Intelligence and Statistics*, Fort Lauderdale, FL, 434–442, 2011b.

B. Li, S. C. Hoi, P. Zhao, and V. Gopalkrishnan. Confidence weighted mean reversion strategy for on-line portfolio selection. In *ACM Transactions on Knowledge Discovery from Data*, 2013.

B. Li and S. C. H. Hoi. On-line portfolio selection with moving average reversion. In *Proceedings of the International Conference on Machine Learning*, Edinburgh, 273–280, 2012.

B. Li, P. Zhao, S. Hoi, and V. Gopalkrishnan. PAMR: Passive–aggressive mean reversion strategy for portfolio selection. *Machine Learning*, 87(2):221–258, 2012.

A. W. Lo. Where do alphas come from? A measure of the value of active investment management. *Journal of Investment Management*, 6:1–29, 2008.

A. W. Lo and A. C. MacKinlay. When are contrarian profits due to stock market overreaction? *Review of Financial Studies*, 3(2):175–205, 1990.

M. S. Lobo, M. Fazel, and S. Boyd. Portfolio optimization with linear and fixed transaction costs. *Annals of Operations Research*, 152(1):341–365, 2007.

J. Loveless, S. Stoikov, and R. Waeber. Online algorithms in high-frequency trading. *Communication of the ACM*, 56(10):50–56, 2013.

C.-J. Lu, T.-S. Lee, and C.-C. Chiu. Financial time series forecasting using independent component analysis and support vector regression. *Decision Support Systems*, 47:115–125, 2009.

D. G. Luenberger. *Investment Science*. New York: Oxford University Press, 1998.

L. C. MacLean, E. O. Thorp, and W. T. Ziemba. *The Kelly Capital Growth Investment Criterion: Theory and Practice*. Volume 3. Singapore: World Scientific, 2011.

M. Magdon-Ismail and A. Atiya. Maximum drawdown. *Risk Magazine*, 10:99–102, 2004.

C. D. Manning and H. Schütze. *Foundations of Statistical Natural Language Processing*. Cambridge, MA: MIT Press, 1999.

D. Maringer. Constrained index tracking under loss aversion using differential evolutionary, natural computing in computational finance. In *Natural Computing in Computational Finance*, A. Brabazon and M. O'Neill (eds.), 7–24. Berlin: Springer, 2008.

H. Markowitz. Portfolio selection. *The Journal of Finance*, 7(1):77–91, 1952.

H. Markowitz. *Portfolio Selection: Efficient Diversification of Investments*. New York: Wiley, 1959.

T. H. Mcinish and R. A. Wood. An analysis of intraday patterns in bid/ask spreads for NYSE stocks. *The Journal of Finance*, 47(2):753–764, 1992.

B. McWilliams and G. Montana. Sparse partial least squares regression for on-line variable selection with multivariate data streams. *Statistical Analysis and Data Mining*, 3(3):170–193, 2010.

N. Meade and G. R. Salkin. Index funds-construction and performance measurement. *The Journal of the Operational Research Society*, 40(10):871–879, 1989.

N. Meade and G. R. Salkin. Developing and maintaining an equity index fund. *The Journal of the Operational Research Society*, 41(7):599–607, 1990.

M. H. Miller. Financial innovation: The last twenty years and the next. *The Journal of Financial and Quantitative Analysis*, 21(4):459–471, 1986.

T. Mitchell. *Machine Learning*. Burr Ridge, IL: McGraw-Hill, 1997.

J. Moody and M. Saffell. Learning to trade via direct reinforcement. *IEEE Transactions on Neural Networks*, 12(4):875–889, 2001.

J. Moody, L. Wu, Y. Liao, and M. Saffell. Performance functions and reinforcement learning for trading systems and portfolios. *Journal of Forecasting*, 17:441–471, 1998.

Y. Nevmyvaka, Y. Feng, and M. S. Kearns. Reinforcement learning for optimized trade execution. In *Proceedings of the International Conference on Machine Learning*, 673–680, 2006.

E. Ordentlich. *Universal Investment and Universal Data Compression*. PhD thesis, Stanford University, 1996.

E. Ordentlich. Encyclopedia of Quantitative Finance, *Universal Portfolios*. Sussex: Wiley, 2010.

E. Ordentlich and T. M. Cover. On-line portfolio selection. In *Proceedings of the Annual Conference on Learning Theory*, Desenzano del Garda, Italy, 310–313, 1996.

E. Ordentlich and T. M. Cover. The cost of achieving the best portfolio in hindsight. *Mathematics of Operations Research*, 23(4):960–982, 1998.

M. F. M. Osborne. Brownian motion in the stock market. *Operations Research*, 7(2): 145–173, 1959.

G. Ottucsák and I. Vajda. An asymptotic analysis of the mean-variance portfolio selection. *Statistics and Decisions*, 25:63–88, 2007.

D. C. Porter. The probability of a trade at the ask: An examination of interday and intraday behavior. *The Journal of Financial and Quantitative Analysis*, 27(2): 209–227, 1992.

J. M. Poterba and L. H. Summers. Mean reversion in stock prices: Evidence and implications. *Journal of Financial Economics*, 22(1):27–59, 1988.

E. Qian, R. Hua, and E. Sorensen. *Quantitative Equity Portfolio Management: Modern Techniques and Applications*. Boca Raton: Chapman & Hall/CRC, 2007.

L. Rabiner and S. Levinson. Isolated and connected word recognition—theory and selected applications. *IEEE Transactions on Communications*, 29(5):621–659, 1981.

T. Rakthanmanon, B. Campana, A. Mueen, G. Batista, B. Westover, Q. Zhu, J. Zakaria, and E. Keogh. Searching and mining trillions of time series subsequences under dynamic time warping. In *Proceedings of the 18th ACM SIGKDD International Conference on Knowledge Discovery and Data Mining*, Beijing, China, 262–270, 2012.

M. R. Reinganum. The anomalous stock market behavior of small firms in January: Empirical tests for tax-loss selling effects. *Journal of Financial Economics*, 12(1):89–104, 1983.

J. Rissanen. A universal data compression system. *IEEE Transactions on Information Theory*, 29(5):656–663, 1983.

H. Sakoe and S. Chiba. Dynamic programming algorithm optimization for spoken word recognition. In *Readings in Speech Recognition*, A. Waibel and K. Lee (eds.), Morgan Kaufmann, San Mateo, 159–165. 1990.

S. Shalev-Shwartz. Online learning and online convex optimization. *Foundations and Trends in Machine Learning*, 4(2):107–194, 2012.

W. F. Sharpe. A simplified model for portfolio analysis. *Management Science*, 9: 277–293, 1963.

W. F. Sharpe. Capital asset prices: A theory of market equilibrium under conditions of risk. *The Journal of Finance*, 19(3):425–442, 1964.

W. F. Sharpe. Mutual fund performance. *The Journal of Business*, 39(1):119, 1966.

W. F. Sharpe. The sharpe ratio. *Journal of Portfolio Management*, 21(1):49–58, 1994.

Y. Singer. Switching portfolios. *International Journal of Neural Systems*, 8(4): 488–495, 1997.

R. Srikant and R. Agrawal. Mining sequential patterns: Generalizations and performance improvements. In *Proceedings of the 5th International Conference of Extending Database Technology*, Avignon, France, 1–17, 1996.

G. Stoltz and G. Lugosi. Internal regret in on-line portfolio selection. *Machine Learning*, 59(1–2):125–159, 2005.

Y. Takano and J.-y. Gotoh. Constant rebalanced portfolio optimization under nonlinear transaction costs. *Asia-Pacific Financial Markets*, 18:191–211, 2011.

N. Taleb. *Fooled by Randomness: The Hidden Role of Chance in Life and in the Markets*. New York: Random House, 2008.

F. E. H. Tay and L. Cao. Application of support vector machines in financial time series forecasting. *Omega*, 29(4):309–317, 2001.

E. O. Thorp. Optimal gambling systems for favorable games. *Review of the International Statistical Institute*, 37(3):273–293, 1969.

E. O. Thorp. Portfolio choice and the Kelly criterion. In *Proceedings of the Business and Economics Section of the American Statistical Association*, Fort Collins, Colorado, 599–619, 1971.

E. O. Thorp. The Kelly criterion in blackjack, sports betting, and the stock market. In *Proceedings of the International Conference on Gambling and Risk Taking*, Montreal, 1997.

R. Tibshirani. Regression shrinkage and selection via the lasso. *Journal of the Royal Statistical Society (Series B)*, 58:267–288, 1996.

J. Ting, T. Fu, and F. Chung. Mining of stock data: Intra- and inter-stock pattern associative classification. *Threshold*, 5(100):5–99, 2006.

E. Tsang, P. Yung, and J. Li. Eddie-automation, a decision support tool for financial forecasting. *Decision Support Systems*, 37:559–565, 2004.

R. S. Tsay. *Analysis of Financial Time Series*. New York: Wiley, 2002.

I. Vajda. Analysis of semi-log-optimal investment strategies. In *Proceedings of Prague Stochastic*, Prague, 2006.

V. Vovk. Derandomizing stochastic prediction strategies. In *Proceedings of Annual Conference on Computational Learning Theory*, Nashville, Tennessee, 32–44, 1997.

V. Vovk. Derandomizing stochastic prediction strategies. *Machine Learning*, 35: 247–282, 1999.

V. Vovk. Competitive on-line statistics. *International Statistical Review/Revue Internationale de Statistique*, 69(2):213–248, 2001.

V. G. Vovk. Aggregating strategies. In *Proceedings of the Annual Conference on Learning Theory*, Rochester, NY, 371–383, 1990.

V. G. Vovk and C. Watkins. Universal portfolio selection. In *Proceedings of the Annual Conference on Learning Theory*, Madison, WI, 12–23, 1998.

X. Wang, A. Mueen, H. Ding, G. Trajcevski, P. Scheuermann, and E. Keogh. Experimental comparison of representation methods and distance measures for time series data. *Data Mining and Knowledge Discovery*, 26(2):275–309, 2013.

R. J. Yan and C. X. Ling. Machine learning for stock selection. In *Proceedings of the ACM SIGKDD International Conference on Knowledge Discovery and Data Mining*, San Jose, CA, 1038–1042, 2007.

H. Yang, Z. Xu, I. King, and M. R. Lyu. Online learning for group lasso. In *Proceedings of the International Conference on Machine Learning*, Haifa, Israel, 1191–1198, 2010.

B.-K. Yi, H. Jagadish, and C. Faloutsos. Efficient retrieval of similar time sequences under time warping. In *Proceedings of the International Conference on Data Engineering*, Orlando, FL, 201–208, 1998.

W. Young. Calmar ratio: A smoother tool. *Futures*, 20(1):40, 1991.

W. Zhang and S. Skiena. *Financial Analysis Using News Data*. Technical report, State University of New York at Stony Brook, 2008.

W. Zhang and S. Skiena. Trading strategies to exploit blog and news sentiment. In *Proceedings of the International AAAI Conference on Weblogs and Social Media*, Atlanta, 375–378, 2010.

Y. Zhou, R. Jin, and S. C. Hoi. Exclusive lasso for multi-task feature selection. In *Proceedings of the International Conference on Artificial Intelligence and Statistics*, Chia Laguna Resort, Sardinia, Italy, 988–995, 2010.

M. Zinkevich. Online convex programming and generalized infinitesimal gradient ascent. In *Proceedings of the International Conference on Machine Learning*, Washington, DC, 928–936, 2003.

H. Zou. The adaptive lasso and its oracle properties. *Journal of the American Statistical Association*, 101:1418–1429, 2006.

Index

Note: Page numbers ending in "f" and "t" refer to figures and tables, respectively.

For further Sales, Overseas and Information please contact our
European agents, CPI/Ferwe medienmanufaktur GmbH Taylor & Francis
Verlag GmbH, Kaufingerstraße 21, 30151 Altenburg, Germany